THE SECRET LIFE OF
CLAMS

Susie—
Here's to all
the Good Times!

Cheers,

Tony

Selected Works by Anthony D. Fredericks

Adult Nonfiction

Walking with Dinosaurs
Horseshoe Crab: Biography of a Survivor
How Long Things Live
Ace Your Teacher Interview
Historical Trails of Eastern Pennsylvania
More Science Adventures with Children's Literature
Desert Dinosaurs
Nonfiction Readers Theatre
Lancaster and Lancaster County

Children's Books

Mountain Night, Mountain Day
Under One Rock
The Tsunami Quilt
Desert Night, Desert Day
A is for Anaconda
Cannibal Animals
Around One Log
I Am the Desert
P is for Prairie Dog
In One Tidepool
Tsunami Man

THE SECRET LIFE OF

CLAMS

*The Mysteries and Magic of
Our Favorite Shellfish*

Anthony D. Fredericks

Skyhorse Publishing

Skyhorse Publishing books may be purchased in bulk at special discounts for sales promotion, corporate gifts, fund-raising, or educational purposes. Special editions can also be created to specifications. For details, contact the Special Sales Department, Skyhorse Publishing, 307 West 36th Street, 11th Floor, New York, NY 10018 or info@skyhorsepublishing.com.

Skyhorse® and Skyhorse Publishing® are registered trademarks of Skyhorse Publishing, Inc.®, a Delaware corporation.

Visit our website at www.skyhorsepublishing.com.

10 9 8 7 6 5 4 3 2 1

Library of Congress Cataloging-in-Publication Data is available on file.

Cover design by Brian Peterson
Cover photo credit Thinkstock

Print ISBN: 978–1–62914–697–3
Ebook ISBN: 978–1–63220–118–8

Printed in China

To the memory of Jerry Ohl—
A heck of a brother-in-law,
A hell of a friend!

Contents

PART THREE: Business and Pleasure

Introduction

I was training to be an electrician. I suppose I got wired the wrong way round somewhere along the line.

—Elvis Presley

IN 1965, A MUSICAL ROMANTIC COMEDY, STARRING Elvis Presley, was released in movie theaters all across America. *Girl Happy* was typical of many beach-party films produced in the sixties—a simple plot, lots of campus hijinks, wild parties, and a loud soundtrack. The story, such as it was, revolved around a group of college students going to Fort Lauderdale, Florida, for their annual spring break. Boy woos girl, boy loses girl, and boy gets girl again takes up the majority of this predictable ninety-six-minute movie.

But, as was the case in all of Elvis's movies, it wasn't the plot that got moviegoers into the theaters, it was his singing. This was Elvis's eighteenth movie and it, too, was packed with lots of rock-and-roll songs, including "Startin' Tonight," "Puppet on a String," and the shake-'em-up dance "Do 'The Clam.'"

"Do 'The Clam'" invited beachcombers to grab whomever was nearest to them and (with the appropriate background of bongo music) to really get the beach a-rocking. Dancers were encouraged to do some turning, some teasing, some hugging and some squeezing—the combination of which constituted the necessary and basic moves of this new dance sensation.

I doubt whether the song's lyrics ever endeared edible bivalves to the throngs of teenagers who flocked to see this film. Nevertheless, this song is one of the few musical numbers to highlight clams—albeit with a less than memorable melody.

CB

I initially grew up in Los Angeles. Then, a family move forty miles south to Newport Beach meant that a significant portion of my southern California childhood was spent body-surfing long cresting waves near the Newport Pier, slipping and sliding over the tidepools at Little Corona, water skiing behind super-charged motorboats in Back Bay, and laughing through endless summer volleyball games on the white sands of Beacon Bay. I also had the insufferable habit of collecting all manner of flotsam and jetsam that invariably washed up on the beach after a summer storm. However, I was particularly fascinated with sea creatures—including the sea anemones and sea stars hidden in the deep recesses of tidepools, the weird invertebrates I had to scrape off the underside of my tiny sailboat, or the remains of a strange fish washed up with the tide. I was also impressed with the occasional shark or two sighted off the Balboa Peninsula as well as the enormous albacore my parents would haul home after their various fishing ventures near Catalina Island.

FAST FACT: Santa Catalina Island has been inhabited for at least 8,000 years. In the 1930s–1950s the Chicago Cubs baseball team held spring training on the island.

Newport Pier.

But it was clams I loved the most. On occasional summer days my parents would send my two sisters and me out with buckets and shovels to dig along the beach fronting our house. We were on the prowl for clams in any size, any shape. And we found them . . . in droves. We invented games: who could find the most, who could find the biggest, or who would fill their bucket first. Then, it was a mad dash back to the house—our shouts and screams echoing down the bay—eventually dumping our caches into the kitchen sink.

Our mother would place a large pot of water on the stove and bring it to a rolling boil. Nearby was a deep pan with two or three sticks of butter slowly melting away into a warm pool of delicious delight. Silverware would be set, large handfuls of napkins would be deposited in the middle of the table, and soon the feast would begin. Our plates crowded with clams, our chins dripping with long rivulets of melted butter, and our laughter punctuating the summer air would make for joyous memories I have kept for more than a half century.

Clams, it seemed, brought out all the good things in life. It was a time for grown-ups and kids to share the stories of our summer and the familial camaraderie that is so much a part of slower days. Occasionally, we would eschew the beachfront clams and travel to one of our favorite seafood eateries. We'd pile into my father's Buick and head over to the Balboa peninsula and The Crab Cooker—a venerable Orange County institution that's been serving seafood delicacies since 1951. Hot platters piled with clams, large crocks of melted butter (with bulky slices of lemon), and a cloth napkin hanging under one's chin made for an unforgettable meal. Occasionally we would tinker with sand dab fillets or planks of halibut, but we always came back to the clams—as appetizers or the main course—those savory, juicy, and oh so succulent clams. There were never enough!

For my wife, the memories were no less sweet. Growing up in northern New Jersey, she and her family would make regular pilgrimages to The Clam Broth House, the iconic seafood restaurant that's been dishing out clams in Hoboken, New Jersey, since 1899. (Jackie Gleason was a regular patron.) There, they would cram themselves into cramped booths, tuck enormous napkins into their shirts, and devour cascading platters of steamers. Conversation was muted, but the culinary satisfaction never was! Like The Crab Cooker, The Clam Broth House has become a gastronomic staple of many childhoods and many family memories over the generations.

The Clam Broth House.

As I grew older I never lost my youthful fascination with the sea (and its denizens), for once saltwater courses through your veins you can never fully divorce yourself from any ocean for very long. Although I now live in a semi-rural region of south central Pennsylvania, I am still drawn to the sea . . . and especially to its food. My wife and I often escape to the bustling shores of New Jersey, the placid beaches of Delaware, the languid coastline of Virginia, the tranquil seashores of Florida, the booming surf of California, the misty shores of Washington and Oregon, and the vibrant volcanic sands of Hawaii on various vacations and journeys—much as Arctic terns are inexorably drawn from their Arctic breeding grounds to their wintering grounds off Antarctica in their annual migration of both body and spirit.

FAST FACT: The Arctic tern has the farthest yearly journey of any bird—up to 56,000 miles round trip.

As a writer of science (and a consumer of bivalves), my seashore journeys often bring me into contact with adventuresome folks—oceanic comrades with a penchant for culinary quests, seaside excursions, or off-the-grid discoveries. I am also honored by the company of marine biologists and conservationists with a keen eye for the mysteries and riddles of the sea. It is with the assistance, support, and expertise of these folks that I am able to maintain my obsession for sea life of every stripe and color, and I can continue my celebration of maritime discoveries (both epicurean and scientific) in myriad tropical waters, warm bays, languid harbors, or distant beaches.

Thus, I was delightfully surprised to learn that the topic of clams was somewhat more complex and diverse than I had been led to believe as a child. In my youth, clams were dinnertime treats; yet, as I was to discover on this research venture, these burrowing critters were somewhat more complex in design, function, and influence than I had previously known. It is those discoveries that I share with you, dear reader. Indeed, for many, this book may well be a first in-depth exploration of these mysterious creatures—their first look at an animal that has beaten the environmental and evolutionary odds and survived for longer than 98 percent of all the animals that have ever been. You will find, as did I, that it is an animal that has influenced civilizations, changed history, become an economic indicator, and been established as a cultural and entertainment icon. Not bad for a bottom-feeding invertebrate!

And so, I invite you, fellow explorer, to join me in this current love affair with a most incredible creature, one who will satisfy your scientific curiosity as well as your gastronomic desires.

Or, as Elvis would say, let's "dig right in and do 'The Clam'."

Prologue

Q: Why don't clams and oysters share their toys with each other?
A: They're two shellfish.

WHEN I WAS AN UNDERGRADUATE STUDENT AT the University of Arizona in the late 1960s, there was (and continues to be) an intense rivalry between us and Arizona State University. This rivalry reached an incredible crescendo during football season, when thousands of students would gather in pep rallies and other pregame celebrations to denounce, criticize, and denigrate the other team in myriad creative ways (some G-rated, some not). We always thought we were better than our rivals (as they did of us), and the day of "The Big Game" was a day like no other. It was the game of all games! It was the defining moment of the entire football season. Irrespective of the win/loss record at that point in the season, the only thing that mattered was beating ASU. Beat them, and it was a great season. Lose, and suffer mightily for the next 364 days. Similar circumstances surround professional sports rivalries.

Rivalries, as you might expect, spill over into our everyday lives as well. We are subjected to rivalries in what we imbibe ("The King of Beers" vs. "Brewed with Pure Rocky Mountain Spring Water"), in what we guzzle (Coke vs. Pepsi), in what we drive (Ford vs. Chevy), and even in the underwear we wear (Jockey vs. Fruit of the Loom).

Those rivalries are also part of our culinary predilections (McDonald's vs. Burger King; Outback vs. LongHorn; Chipotle vs. Qdoba). Some of those rivalries are the result of sustained and ubiquitous advertising; others are manufactured by competing epicurean interests bent on establishing their specific food items (or menu) as better than anything else procured, cooked, or served.

So it is with clams and oysters.

There are those who will dig their heels firmly in the mud (or a sandy beach at low tide) and argue that clams have got oysters beat six ways to Sunday. There are others who will hold their breath, stomp their feet, and demand that oysters are the *crème de la crème* of seafood and anyone who says differently doesn't know their elbow from the ever-present hole in the ground. The arguments will escalate, the verbiage will get intense, and the accusations about someone's ancestors will reach a fever pitch. It's apparent things have gotten out of hand.

I'll not be responsible for inciting or propelling those clam vs. oyster disagreements. For me, it's the proverbial "apples and oranges" argument. Some prefer one kind of shellfish; others prefer the other. Is one better than the other? I don't know. Is one preferred more than the other—perhaps yes. I happen to be an aficionado of clams. I have tried oysters and found them wanting. Maybe it's the "slime factor" or the fact that they are so expensive in the restaurants I visit. Maybe it's because their shells look . . . well . . . look so misshapen. Maybe it's because they are the ocean's great filters, passing all manner of microscopic matter and human detritus through their systems. Or, maybe it's just because I grew up with clams in my front yard and on my dinner table and in the coastal towns and villages I frequent that they have become so much a part of my existence—and so much a part of my menus.

It's not that I dislike oysters, it's simply that I don't prefer them. If given a choice at a restaurant, I'll always select my shellfish alphabetically. I've eaten clams in a hundred different recipes and in a hundred different restaurants. I'm comfortable with them, they are my friends, and they make me (gastronomically) happy. Perhaps, some day, oysters might do the same . . . but, after considerably more than six decades of clam companionship, I'm inclined to stick with what I know. Clams are my friends.

I am hoping that you, too, are an enthusiastic aficionado and diehard supporter of clams (a reason, I suppose, why you are reading this book). I hope they are your culinary cohorts, constant colleagues, and compatible compatriots—epicurean delights that have graced your table in innumerable ways. I'm hoping you are eager to learn about their character, their history, their lifestyles, and their influence. However, should you still be on the fence about your shellfish predilections, perhaps the following list may help you decide the debate (indeed, the rivalry) once and for all:

- Clams are bilaterally symmetrical; oysters are, well, odd.
- Elvis sang about clams; he never sang about oysters.
- A quite naked Venus (the goddess of sex and fertility) has been depicted on a clamshell by a famous Renaissance painter (Botticelli); oyster shells haven't had the privilege of supporting any goddesses, naked or otherwise.
- Clams can be found worldwide; oysters are often associated with the United States.
- Clams have more iron than liver; oysters have less iron than eggs.
- Clams are economical (e.g., fried clams); oysters are expensive (e.g., Oysters Rockefeller).
- Clams are proud of *both* their shells; oysters would rather be served on the half-shell.
- Clams will have a positive effect on your libido; oysters will too . . . but do you really want to eat something cold and slimy before a night of incredible lovemaking?

PART I

History and Lore

Chapter 1

Leonardo and Ancient Clam History

What's past is prologue.

—William Shakespeare, *The Tempest*

WHEN I WAS GROWING UP, ONE OF MY CHILD-hood heroes was Davy Crockett. I was enamored of his brav-ery, his ability to set a course of action and stick to it, and of his determination to face problems head on. I was an ardent fan of the TV series based on his life that ran 1954–1955 and starred Fess Parker.* In addition, I made absolutely sure I had a Davy Crockett lunch pail, an offi-cial Davy Crockett shirt (with authentic fringe), and the requisite coonskin hat (which, if I had kept it, would now be worth about a gazillion dollars on eBay). In so many ways, Davy Crocket epitomized what I believed a hero was—fearless, determined, and honest.

Over the years, my heroes have evolved. I've admired many different women and men—not only for who they are, but also for what they believe or think. As I've grown older (and, hopefully, wiser) I've embraced intel-lectual heroes more than the action heroes of my youth. I've looked to art-ists, explorers, scientists, and inventors for their dynamic leadership, innate sense of curiosity, limitless imagination, and infectious creative spirit. I

---

* Fess Parker would later achieve fame (and fortune) as a celebrated vintner and owner of the Fess Parker Winery in Los Olivos, California. Thus, I can continue to celebrate one of my childhood heroes; however, now with a glass of dry Riesling.

admire those who buck the status quo (intellectually speaking), those who think well "outside the box," and those who chart thoughtful investigations far beyond their contemporaries.

Take a few moments and think of some of your heroes—not the Supermen or Wonder Women of your youth, but rather those people, living or dead, who have shaped (or are in the process of shaping) who you are as a fully cognitive person. Who makes you think? Who sparks your creativity or stimulates your curiosity? Who asks the questions no one else is asking? Who are the intellectuals you admire most?

In *The Book of Genius*, Tony Buzan and Raymond Keene apply some very objective criteria in order to rank the greatest thinkers of history. Using categories such as "originality," "versatility," "dominance-in-field," "universality-of-vision," and "strength and energy," they present the following list as their "Top Ten" geniuses of all time:

10. Albert Einstein*
9. Phidias (architect of Athens)
8. Alexander the Great
7. Thomas Jefferson
6. Sir Isaac Newton
5. Michelangelo
4. Johann Wolfgang von Goethe (German writer and politician)
3. The Great Pyramid Builders
2. William Shakespeare

According to the exhaustive research conducted by Buzan and Keene, the greatest thinker, the greatest genius (#1), of all time was Leonardo da Vinci (1452–1519).

By any and all accounts, Leonardo was the ideal Renaissance man, or *uomo universale*, a well-rounded, thoughtful, and inquisitive personality comfortable with both intellectual rigor and social graces. One need only look at a partial listing of some of Leonardo's accomplishments and discoveries to appreciate the extent of his genius. Consider the following:

• As an artist, he painted two of the most famous masterpieces in the world, *Mona Lisa* and *The Last Supper*, along with the world's most famous drawing: *Vitruvian Man* (which has been universally duplicated

---

* On the wall of my office hangs the following Albert Einstein quote: "The difference between stupidity and genius is that genius has its limits."

on items as diverse as coinage, biology textbooks, posters, and the T-shirts worn by some of my undergraduate students).

- As an architect, he designed the cathedrals in Milan and Pavia.
- As an inventor, he created plans for a flying machine, a helicopter, the bicycle, a snorkel, a horizontal waterwheel, the viola organista (a musical instrument), an adding machine, canal locks, an olive press, folding furniture, and a water-powered alarm clock.
- As a student of maps, he pioneered the use of cartographic perspective and map-making accuracy.
- As a mathematician, he studied linear perspective and promoted geometry.
- As an engineer, he drew up plans for an armored tank, machine gun, guided missile, double-hulled boat, and the submarine.
- As a scientist, he pioneered the study of comparative anatomy and modern botanical science, discovered the science of dendrochronology, wrote about the nature of light, observed geotropism in plants, outlined the human body, and noted that the earth was not in the center of the universe.*

> **FAST FACT:** Born on April 15, 1452, in Vinci, Italy, Leonardo da Vinci was the illegitimate son of a twenty-five-year-old notary, Piero da Vinci, and Caterina, a peasant girl. His parents married other people, and eventually Leonardo wound up with seventeen half-brothers and half-sisters.

Leonardo da Vinci, Clams, and a Biblical Flood

Leonardo may well be the "prototype" for today's scientist—one who effectively and efficiently utilizes seven essential scientific processes to explore and understand the surrounding world: observing, classifying, measuring, inferring, predicting, communicating, and experimenting. That Leonardo was able to use these processes in a coordinated and effective system of thinking and imagining is a testament not only to his genius but also to

---

* A survey (of twenty-two hundred participants) conducted by the National Science Foundation in 2014 found that fully 25 percent of Americans still believe that the sun orbits the earth—a fact disproved way back in the sixteenth century by Nicolaus Copernicus.

the array of scientific discoveries and inventions he made throughout his life—including a revelation involving clams!

You see, Leonardo was known as a copious note-taker. In his lifetime, he produced nearly thirty different scientific journals, each filled with drawings, notes, descriptions, and observations of the natural world. His best-known journal, the *Codex Leicester*, is a collection of eighteen sheets of paper, all folded in half and written on both sides. This seventy-two-page document is a random assortment of scientific writings about (among other things) astronomy, the movement of water, geology, the luminosity of the moon, geology, and paleontology.

FAST FACT: The *Codex Leicester* was sold to Bill Gates by Christie's Auction House in New York on November 11, 1994, for $30,802,500. (Yes, you read that amount correctly!)

A significant portion of the *Codex* centers on a refutation of current (at the time) thinking regarding the reasons why clam shells (and their fossils) were found near the tops of European mountains. How did clams come to live at the tops of mountains when they were, most certainly, marine creatures whose natural habitat was the bottom of the sea? How clamshells and clamshell fossils were found at great elevations was a source of constant speculation, guessing, and outright conjecture by the scientists of the day—a mechanism of suppositions not unlike a random assembly of children's party balloons flying off in all directions during a spring breeze.

The persistent mystery was, how did fossil shells (and the remains of other former marine organisms) get caught in strata several thousand feet above current sea level? Thus, based purely on rampant and unsubstantiated guesswork, the scientists of the time concluded that the deposition of these shells was due to the high waters and violent currents of Noah's flood. The answer, they insisted, was in religion, not science.

Leonardo ridiculed this theory in the *Codex* by noting that the accumulation of clam and marine fossils in several different strata confirms the idea that their deposition occurred at different times in the history of the earth. He supported this conclusion with a classic piece of scientific observation (most obviously ignored by his contemporaries) regarding the position of clamshells:

> And we find the [clams] together in very large families, among which some may be seen with their shells still joined together,

which serves to indicate that they were left there by the sea and that they were still living.

Leonardo also inferred that the clamshells were subjected to extensive movement after the death of the organisms:

> In such locality there was a sea beach, where the shells were all cast up broken and divided and never in pairs as they are found in the sea when alive, with two valves which form a covering the one to the other.

One of the major themes of the *Codex* was the explicit refutation and denouncement of Noah's flood as a significant cause of fossil deposition. Two major sections of the document are titled "Of the Flood and of marine shells" and "Refutation of such as say that the shells were carried a distance of many days' journey from the sea by reason of the Deluge."*

Leonardo further attacked a concurrent theory that fossils were not, in fact, the remains of ancient organisms, but rather that they "grew" (as a manifestation of some unknown plastic or cosmic force) within rocks precisely mimicking living creatures. This belief, otherwise known as Neoplatonic theory, although widely embraced by scientists and laypeople alike, could not explain why these apparently inorganic objects were not found in all strata, rather than just in those strata with evidence of oceanic remains. Leonardo was quite adamant (and very livid) about the ridiculousness and stupidity of this "scientific" precept:

> And if you should say that these shells have been and still constantly are being created in such places as these by the nature of the locality and through the potency of the heavens in those spots, such an opinion cannot exist in brains possessed of any extensive powers of reasoning because the years of their growth are numbered upon the outer coverings of their shells [the age of a fossil clamshell, the science of sclerochronology, can often be inferred from growth rings on the shell]; and both small and large ones may be seen, and these would not have grown without feeding or feed without movement, and here [that is, in solid rock] they would not be able to move . . . . Ignoramuses maintain that nature or the heavens have created [fossils] in these places through celestial influences.

---

* Crafting pithy titles was not, apparently, one of Leonardo's strong suits.

Leonardo's observations of clamshells on European mountaintops and the resulting condemnation (nay, damnation) of medieval paleontological thought further cement both his scientific legacy as well as his genius. That bivalves were to play a significant role in the expansion of scientific thought (and the refutation of conjecture and supposition as "fact" or "reason") cannot be undervalued or underappreciated. Clams helped one of the world's greatest thinkers establish a cogent and defensible argument that modern paleontologists still embrace. That is reason enough to celebrate clams' prominence in other arenas of scientific discovery.

The Evolution of Clams

Now, just for a moment, let's consider a number . . . a very large number: 510,000,000.

I think we can agree that 510 million is a very large number—particularly when we apply that figure to time, as in 510 million years.

FAST FACT: 510 million years is equal to:

- 6.1 billion months
- 26.5 billion weeks
- 186.1 billion days
- 1,489.2 billion consecutive National Football League games (averaging three hours each)

Now, let's take a journey—a very brief narrative journey back in time—to examine some of the early "events" that took place on this ever-evolving planet. About 4.55 billion years ago our early Earth, such as it was, consisted of nothing more than ice and rock particles swirling around the young sun. Then, a cosmic force caused them to smash together to create the planet or proto-planet (what we now know as "The Big Bang Theory"—the real one, not the TV series). Then, the planet's surface slowly cooled and for the next 700 million years or so, ours was a completely alien world without water, without oxygen, and certainly without life.

The first part of our new planet's life was known as the Precambrian era, an enormous chunk of time that makes up about seven-eighths of the Earth's history, an expanse of approximately four thousand million years.

As you might imagine, the length of the Precambrian era enabled several important biological, geological, and physical events to take place (albeit, very slowly). These included the rise and movement of the first tectonic plates, the evolution of eukaryotic cells (the cells that make up all animals, plants, fungi, and protists), and the infusion of oxygen into the atmosphere. It was only toward the end of the Precambrian era that complex multicellular organisms, including soft-bodied marine creatures, began to appear.

FAST FACT: Here's something that will demonstrate the incredible length of the Precambrian era: Get (or imagine) a two-by-four piece of lumber from your local lumber yard. Cut it to a length of six feet. That six feet represents the entire extent of earth's history (from the Big Bang to the present day). Take a piece of sandpaper and sand one end of your two by four for about fifteen seconds. The amount of sawdust you just created represents the total time humans have been on the earth. Now, use a saw and cut about eight inches off one end of the two by four. The remaining length of your two by four (sixty-four inches) represents the length of the Precambrian era.

After the Precambrian, the next era in the history of the Earth is known as the Paleozoic (542–251 million years ago). This era (also referred to as "The Time of Early Life") was one of tremendous and incredible growth in the number, diversity, and complexity of animal life. The Paleozoic era comprised six separate time frames, or periods, as follows:

- Cambrian: 542–488 million years ago
- Ordovician: 488–444 million years ago
- Silurian: 444–416 million years ago
- Devonian: 416–359 million years ago
- Carboniferous: 359–299 million years ago
- Permian: 299–251 million years ago

FAST FACT: At the beginning of the Paleozoic era (the Cambrian Period), today's western coast of North America ran east-west along the Equator, while Africa was at the South Pole. What we now know as Europe was essentially an island located below the equator, two-thirds of the way to the South Pole.

It is the Cambrian Period that should capture our attention, at least in a book about clams. It was during this time frame that several significant biological events took place. These included the following:

- Animals with hard skeletons first appeared.
- Plants had still not evolved yet—essentially the terrestrial world was devoid of any vegetation.
- Small, jawless, armored fish (known as ostracoderms) first appeared. These were the first marine creatures to use gills for respiration.
- Trilobites, small marine creatures, were the first to curl up into a ball—a most effective defensive move still being practiced by several other animals today (i.e., pill bugs, armadillos, pangolins).
- Clams show up.

Scientific research has documented the earliest clam fossils in rocks dating to the middle of the Cambrian period, or about 510 million years ago. Later on, during the Devonian Period (416–359 million years ago), clams (and their relatives) became increasingly abundant. However, they really took off following the massive extinction at the close of the Permian Period (251 million years ago). Interestingly, this extinction affected many groups of organisms in many different environments, but it affected marine communities the most by far, causing the extinction of most of the marine invertebrates of the time—with the exception of our friends, the clams.

Today, modern clams live in a variety of marine and freshwater environments, from shallow tidal waters near shore to great depths in the world's oceans. Interestingly, ancient clam fossils indicate that clams have occupied most of these same environments for more than 450 million years. Imagine living in the same community, same neighborhood, and same region for that period of time.

Now, if you were to look at some of those ancient fossil clams, you would note that they are incredibly small. Indeed, most Cambrian species are especially tiny; in fact, they are just large enough to see without magnification. However, over time, larger and larger species evolved. The largest, inoceramid clams, eventually grew to be as much as six feet in diameter. Consider, if you will, a clam twice as large as the tires on your car and you get a good idea of just how enormous those clams were.

Now extinct, inoceramid clams lived in colonies in the shallow seas that once covered much of North America during the Cretaceous Period (from about 145–65 million years ago). They have been preserved in great numbers in the rocks of the Niobrara Chalk, a geologic formation laid down 87–82 million years ago. The chalk formed from the accumulation of billions of microorganisms living in what once was the Western Interior Seaway, an area that underlies much of the Great Plains region in the United States and Canada.

Scientists have noticed that some of these huge fossil clams are covered with encrusting oysters and other organisms (Cretaceous hitchhikers, perhaps). Other inoceramids have been discovered with a variety of fish fossils lodged between their shells, indicating that the fish may have used the giant clams as safe havens or convenient feeding places.

> **FAST FACT:** Several species of inoceramids have been found in the Late Cretaceous rocks of Kansas. Amazingly, they also produced pearls (which have also been recovered).

It is likely you have seen many of these ancient clams (and some of their marine neighbors)—albeit in ways and in locations you might find surprising. For example, if you were to walk around certain sections of New York City and kept your eyes open, you'd encounter the remains of ancient sea creatures trapped inside the facades, storefronts, and soaring stone walls of some of the most luxurious buildings in the city. For example, in the façade of the Sherry-Netherlands Hotel (781 Fifth Avenue) are cross sections of snails, tiny scallop shells, and oversize shells all encased in a stony tomb more than 40 million years old. Cast your eyes at Tiffany's window (Fifth Avenue and 57th Street) and you're likely to discover ancient sea lilies embedded in the surrounding stone, their "home" for more than 100 million years. In the marble of the Tishman Building (666 Fifth Avenue), in full view of any passersby, are the remnants of ancient relatives of the chambered nautilus, striking marine creatures that swam through prehistoric seas for some 365 million years. The cornices of Saks Fifth Avenue (611 Fifth Avenue) house a section of coral reef that once lay under a vast and ancient sea covering almost all of the American Midwest millions of years ago.

Also awaiting your inspection are several "Big Apple" buildings constructed of limestone from northeastern Italy. Sprinkled throughout the walls and decorative adornments of these buildings are the distinctive impressions and fossil remains of ancient clams known as rudists. You see, rudists were a very common marine heterodont bivalve that arose in the late Jurassic and spread prolifically during the Cretaceous period. They are significant, due in large measure to their role as massive reef builders throughout the Tethys Ocean.* As shallow water inhabitants, rudists were known as epifaunal organisms, meaning they were usually attached to the sea floor sediment.

Cretaceous rudist bivalves from the Oman Mountains. (Scale bar is ten millimeters.)

Current thinking is that the salinity and temperature of Cretaceous seas may have been responsible for the environmental success of rudists. Specifically, tropical waters during this time were considerably more saline and considerably warmer than today's equatorial seas. As a result, rudists absolutely flourished in these waters.

FAST FACT: Rudist reefs were sometimes hundreds of feet tall and often ran for hundreds of miles along the continental shelves. At one point in time, a rudist reef followed the curve of the North American coastline from the present-day Maritime Provinces down to the Gulf of Mexico.

---

* This former ocean separated the supercontinent of Laurasia in the north from Gondwana in the south for much of the Mesozoic era (251 to 65.5 million years ago).

Early rudists, those of the Jurassic period, were elongated, with both valves (shells) being similarly shaped—often pipe- or stake-shaped. By contrast, Cretaceous rudists were odd-shaped. Unlike the two shells (or valves) of similar size and shape evidenced by either their predecessors or today's bivalves, the Cretaceous rudists had dissimilar valves. The lower valve was typically large and conical in shape and was attached to neighboring rudists or the seafloor. It was this valve that served as the clam's living space. The smaller valve (usually on top) formed a type of lid for the organism. This smaller valve was sometimes flat and simple, sometimes low and conical in form, occasionally spiral, and in rare cases star-shaped.

Today, there are no known direct descendants of the rudists. The Cretaceous-Paleogene extinction event (K-Pg) effectively wiped out all the rudists 66 million years ago (as it also did with all nonavian dinosaurs). Rudist fossils, however, can be found throughout the equatorial latitudes, generally in the areas comprising the Mediterranean, the Middle East, the Caribbean, and Southeast Asia. And their remains are forever preserved in the facades of several soaring structures looming over the urban landscape.

Early Clams Discovered in the Desert

Just for a moment, let's depart that urban environment and travel some 5,713 miles (via our own private jet) to the Negev Desert in southern Israel. This may seem a strange place to go clamming, but a quarter-billion years ago, this region was a shallow equatorial seacoast bordering the supercontinent of Pangaea.* Eventually the ocean receded and Pangaea fractured, its pieces drifting into the familiar configurations of present-day continents.

However, over time, fissuring and erosion also uncovered a stretch of this ancient seabed in an area called Makhtesh Ramon, a geological feature that's considered Israel's Grand Canyon. (Makhtesh is Hebrew for "crater," though the deep channel resulted from river drainage, not an astronomical impact.). On a visit to Israel in 2007, geologist Mark Wilson (from the College of Wooster in Wooster, Ohio) collected about a dozen unbroken clam fossils from a yellowish, sandy mound in the Makhtesh Ramon. Intrigued by what they might be, he sent the new samples, ranging from babies to full-grown adults, to a colleague in Texas.

---

* Pangaea was an enormous supercontinent that existed during the late Paleozoic and early Mesozoic eras. It formed approximately 300 million years ago and began to break up into today's present continents about 100 million years later.

After a period of intensive research and laboratory investigation, the pair of scientists concluded that the strange-looking clams represented something new. It wasn't just a novel species or an undescribed genus (the next-higher level of biological classification). Wilson's find was actually the discovery of an entirely new clam family. In the world of invertebrate paleontology, that doesn't happen often. Wilson christened the clam family the ramonalinids, incorporating the name of the locality where they were found.

What was most interesting about these creatures was that the ramonalinids were an odd and apparently isolated species of ancient clam. As far as anyone knows, they only lived in a single coastal alcove of the vast prehistoric Tethys Sea. The ramonalinids formed a continuous mound of mud that accumulated over hundreds of thousands of years, eventually reaching a height of nearly twenty feet.

"There were thousands of them in this one place, for a couple of miles," Wilson said. "It was probably an environmental zone that was just right for them. But they're not known anywhere else in the world. They seem to be a lineage that was exploring various ecological niches in which they had little competition."

Most modern clams bury themselves in sea-bottom sediment to hide from snails, crabs, whelks, and sea stars, who like to crack their armor or suck up their innards (more about this gruesome digestive activity in chapter 6). Ramonalinid clams eschewed subterranean habitats and not only sat on top of the seabed, they perched on edge, similar to how you place dinner plates in your dishwasher. To maintain this upright posture, ramonalinids modified a part of their shell, forming a large flat surface that functioned as sort of an underwater snowshoe on which they balanced themselves.

The ramonalinids' two hinged shell halves were nearly an inch thick, and they didn't fully close, probably by design. The gap would have allowed the clams' inner tissue, or mantle, to loll outside, where sunlight could reach it. That way, scientists think, algae and other photosynthetic microbes living in the clams' mantle could convert sunlight into nutrients, which the clams could use as supplemental food. The metabolic boost from these tiny symbionts likely helped the clams make and maintain their massive shells. (We'll see this again in a more modern clam—the giant clam of the South Pacific—which we will meet in chapter 3.)

While this surface-dwelling, sunbathing lifestyle had benefits, it also carried risks. "To be a big, meaty organism and just sit on the seafloor was

dangerous," Wilson said. Though the ramonalinids' chunky armor probably helped discourage attackers, it didn't ensure long-term survival. Perhaps new predators with new attack strategies appeared, or further environmental changes proved too much to overcome. The ancient clam family appears to have no modern descendants; in a word, they became an evolutionary deadend.

"All we know is that there's this moment of time where they appear, and then disappear," Wilson said.

Finding Weather Patterns in Clamshells

Now, let's reboard our transoceanic private jet and travel 7,507 miles in a southwesterly direction to the fifth largest continent in the world, Antarctica. In one of the most fascinating research studies in modern times, an international team of researchers has been investigating 50-million-year-old clamshells from this vast and forbidding environment. What they have discovered has rewritten the book on climate change. Specifically, they found that the growth rings of ancient clam fossils demonstrate a predictable climate rhythm over the South Pacific, especially during the last prolonged interglacial phase of Earth's history (which ended about ten thousand years ago). This rhythm resembles the present-day interplay of El Niño (a band of warm ocean water temperatures that periodically develops off the western coast of South America) and La Niña (the counterpart of El Niño, in which the sea surface temperature of the eastern central Atlantic Ocean is lowered by three to five degrees Celsius).

> **FAST FACT:** During the Cambrian Period (541–485 million years ago), western Antarctica was partially in the northern hemisphere. Eastern Antarctica was at the equator, surrounded by tropical seas. Then, more than 170 million years ago, Antarctica became part of the supercontinent Gondwana. Over time, Gondwana gradually broke apart and Antarctica (as we know it today) was formed approximately 25 million years ago.

This climatic phenomenon still changes regularly from its cold phase (La Niña) to the warm phase (El Niño) and back again. Thomas Brey, a biologist at the Alfred Wegener* Institute for Polar and Marine Research (in Germany), and his colleagues have succeeded in verifying this pattern as an ancient one. They have done so by observing the growth patterns of Antarctica fossil clams from the early Eocene epoch (about 56–34 million years ago).

Brey and his colleagues investigated the shells from the bivalve species *Cucullaea raea* and *Eurhomalea antarctica*, shells that are 50 million years old. "Like trees, clams form growth rings. We measured their width and examined them for growth rhythms," stated Brey.

Whether clams grow ultimately depends on two significant factors—the availability of food and environmental warmth. "That means the change from 'good' and 'poor' environmental conditions at that time is still reflected in the width of the growth rings we find today," said Brey. "And as we were able to show, this change took place in the same three to six year rhythm we are familiar with in connection with El Niño and La Niña today."

The shells are a real piece of paleontological luck. "To verify El Niño, we need climate archives that cover the largest possible period year by year. Back then, clams lived for up to 120 years. This is a good basis for our work," he said.

To examine the significance of the growth rings of clams, the researchers compared their measurement results with current El Niño data as well as with the El Niño–like fluctuations produced by a climate model of the Eocene. The result? All patterns correspond. "Our results are a strong indication that an [El Niño] phenomenon which fluctuated between warm and cold phases also existed in the warm Eocene," concluded Brey.

What the Antarctic clam fossils reveal is that climate tends to be cyclical in nature. Even more important is the fact that those cycles are permanently recorded by organisms that have been around for far longer than the considerably more intelligent beings studying them. While we might think of clams as innocent bystanders to the world's evolutionary

---

* Alfred Wegener was the German scientist who first proposed (in 1912) the theory of continental drift—the theory that the continents were slowly drifting around the earth. Now referred to as tectonic plate theory, his views were not readily accepted until the late 1950s. (One prominent scientist in the 1920s emphatically stated that Wegener's ideas were "Utter, damn rot!") Today those ideas are established scientific fact, and Alfred Wegener is now recognized as the founding father of one of the major scientific revolutions of the twentieth century.

and climatic changes, they have certainly proven themselves as invaluable "stenographers" of the cycles of time.

After traipsing around the frigid environment of our southernmost continent, it is now time to return to a considerably warmer clime. We'll hop back onto our supersonic private jet and travel back to the Middle East. It is here that another significant bit of clam news has been unearthed—one that will put a fitting exclamation point to this chapter on the early history of our favorite shellfish.

Clams of the Red Sea and Early Human Migration

In 2008 an international team of researchers discovered a new species of giant clam (*Tridacna costata*) in an area surrounding the Red Sea. The fossils indicate that giant clams once dominated elevated reef terraces throughout the area. As a result, they were a ready source of food for the people who lived around the Red Sea. Unfortunately, it seems as though they began to be overharvested as early as 125,000 years ago, leading to the apparent near collapse of the species. But what is critical here is that those same researchers infer that this particular mollusk may hold clues to how and why humans migrated out of Africa more than a hundred thousand years ago.

FAST FACT: The Red Sea is the world's northernmost tropical sea.

The scientists' findings feed speculation that modern humans migrating out of Africa into the Red Sea region 110,000 years ago were motivated in part by disappearing seafood, specifically clams, which may be one of the earliest known examples of marine overharvesting. The discovery also illustrates humans' ancient dependence on the natural environment, said the study's lead author, Claudio Richter, a marine ecologist at the Alfred Wegener Institute.

"The further we got back in time, the more *T. costata* we found. Also the shells were much larger than at present," Richter said. He also stated, "The new giant clam is found exclusively in very shallow waters within easy reach of humans, which makes it much more vulnerable to overfishing than the other species." "The striking loss of large specimens is a smoking gun indicating over-harvesting."

"That theory tracks with history," said team member Marc Kochzius, who conducted genetic analysis on the new species. "The decline of *T. costata* coincided with humans migrating out of Africa," he said. "We propose that giant clams, and especially *Tridacna costada*, were a valuable food resource, which was rather easy to collect on the shallow reef flat," he said.

That clams have been around for more than 500 million years, have diversified into so many different species, and exist in almost every corner of the world is a testament to their persistence as well as their evolutionary success. While we may often consider the lowly clam to be an innocent bystander to the events of history, here is at least one instance when these enigmatic creatures may have effectively influenced the course of human history. That they have determined the survival and ultimate migrations of humans across vast geographic distances is amazing in its own right; that they may have been influential in the fate of other cultures and other peoples will be left to additional explorations by paleontologists, anthropologists, and marine biologists in the future. At this point, however, it seems fair to state that clams may have influenced the course of human events in ways far more significant than how we can simmer them with some cream, butter, and potatoes to create a good-tasting chowder.

Suffice it to say, ancient clams have left us with many secrets to decipher. Let's take a look at some of those secrets in a little more detail.

Chapter 2

Secrets in the Middens

One man's trash is another man's treasure.

—popular saying

IF YOU'VE EVER OWNED A CAT, YOU KNOW THEY often have personalities that are just a little bit short of strange or, shall we say, weird. They have a tendency to do weird things, act in weird ways, and just get plain weird—especially when they crave attention (which is always!). Perhaps that is part of their allure, or perhaps that is just part of their natural ability to weasel (another interesting animal) their way into our lives and our hearts.

When my children were young, we had a cat with a most distinctive habit. It would constantly steal bits of paper, scraps of aluminum foil, a few chicken bones, old batteries, plastic kitchen utensils, and small toys and hide them in secret caches in corners of the living room or kitchen. These piles would be added to over the weeks—an assembly of potpourri, scraps, and the detritus of our life that had no order or logic. If it looked interesting and was easily portable, then it quickly became part of one of his caches. Every month or so, we would have to move the couch or reposition the kitchen hutch to locate and clean out his miniature storehouses of accumulated debris. His real name was Walter, but everyone in the family affectionately called him "Rat" simply because he had the habits of the common pack rat. Rat never met a bit of trash he didn't like!

Pack rats (often referred to as wood rats) can be found throughout the United States. The twenty-two species are quite diverse, living in ecosystems as varied as the deserts of the American southwest, to the deciduous forests of the East coast, to the cold, rocky habitats of the mountain states, and to the juniper woodlands of the western United States. Pack rats often build complex houses or dens containing several nest chambers, food storehouses, and, of course, debris piles.

Pack rats are collectors of the first order, and it is these collections that are particularly interesting. Anything and everything is "fair game" for a pack rat (as it was for our "Rat"). Most of the materials gathered by a pack rat will come from an area extending out to several dozen yards from its nest. These may include items such as sticks, trash, leaves, paper, metal objects (especially coins), branches, wire, foil, rocks, animal dung, or anything else that looks interesting and that can be easily transported from one place to another. If a pack rat can move it, it usually gets moved.

These debris collections are referred to as "middens." To ensure that a pack rat's "belongings" are his (or hers), and his (or hers) alone, he or she will frequently urinate on the debris pile. In so doing, he (or she) has effectively labeled or personalized that midden, preventing potential interlopers from scrambling away with items that don't "belong" to them.*

As the urine dries, sugar and liquids evaporate. This material is now known as *amberat*—a thick resinous substance that tends to cement midden materials together. Over time, the amberat dramatically slows the rate at which those materials decay or decompose. This, in combination with an often arid climate in the places where pack rats frequently live and their predilection of creating middens inside caves or under rocky overhangs, helps preserve those "urinated" debris over time. Interestingly, several pack rat middens, particularly in the American southwest, have been preserved for more than 50,000 years. The discovery of these ancient middens often sets archaeologists into a (scientific) frenzy for two primary reasons. First, the middens reveal a great deal of information about past climatic conditions (through an analysis of woody materials in the middens). Second, these middens often contain very convincing evidence about the prominent plant species that existed thousands of years ago in a particular geographic region. In so many ways, pack rat middens are "crystallized time capsules" of life from long ago.

---

* Apparently, pack rats aren't the only creatures to pee on their belongings. As you may know, urine has a high nitrogen content. Thus, it can be used to harden metal. In 1983, the BMW Formula 1 engineering team was trying to determine how to deal with the high pressures attained by a turbocharged racing engine. That is, how could they strengthen their cars' engine blocks? The solution: they urinated on the engine blocks. (P. S. Don't try this with the family car.)

FAST FACT: Pack rats aren't the only animals that create middens. Octopi do, too! Octopus middens are assemblies of debris the octopus piles up to conceal the entrance of its den. Octopus middens are commonly made of rocks, shells, and the bones of prey, although they also may contain anything the octopus finds that it can move (sounds familiar). In contrast to pack rats, however, octopi do not urinate on their stuff.

Middens in Archaeology

Now, imagine doing the following:

Go into your kitchen and grab your trashcan—the one under the sink or next to the stove. Cover your kitchen table with several layers of newspaper and then dump your trash can out onto the table. What you will see before you is evidence of the events, experiences, and traditions your family has dealt with in the past several days (depending on how often you empty your can). Paw through the debris and you'll discover evidence of the meals you've eaten, the type of mail you've received (particularly if you forgot to put your junk mail in the recycling bin), food packaging, discarded meals, Band-Aids®, broken shoe laces, as well as varied family items considered waste.

Now, let's take that activity to the extreme.

Take all your trash from *an entire year*, load it into the largest rental truck you can find, and drive it over to your local high school football field. Now, dump all that trash onto the field from one end zone to the other. Gather your family members together and take some time (a whole day, for example) to examine all the bits of trash and garbage and waste you have tossed out in the past year. You will undoubtedly discover some of your dietary trends, evidence of your financial habits, some of the important things in your daily living ("to do" lists), how you communicate with the outside world (i.e., phone records, Internet service), trips you've taken or trips you wish you'd taken, as well as a variety of other habits, customs, and practices. In a sense, there would be a fairly accurate archaeological record of your family right there on the football field. (By the way, be sure to clean up the field before you leave.)

In many ways, the trash people leave behind offer those who go through that trash a pretty good picture of what those people did, what they ate, and what they valued. Imagine the excitement, then, when archaeologists discover an ancient trash heap, an accumulation of the debris—not just of one family—but of an entire village or an entire community. Imagine the archaeological exhilaration when that pile of trash is determined to be not just a few years old but thousands, if not tens of thousands, of years old. Those archaeologists would be literally dancing in the street. Such trash piles have the potential of yielding incredible amounts of scientific information and offering significant and revealing clues about a culture or a people.

As "producers of trash," modern humans are not unique. In fact, ever

> **FAST FACT:** In a lifetime, the average American will personally throw away six hundred times his or her body weight, or approximately 4.6 pounds of trash per day.

since recorded history, ancient humans have been creating (and disposing of) various forms of trash or garbage. In the past, our ancestors may have thrown their trash anywhere they wanted. (Environmental concerns were not a serious issue.) In fact, it was not unusual, especially in medieval European cities, for folks to throw their garbage right out the window and into the street. The gentlemanly custom of a man walking on the left side of his female companion when strolling down a street was so that any garbage inadvertently (or intentionally) thrown out a second story window would land on him rather than on his companion. Please consider that people also threw other things out their windows, including (if you can believe it) the contents of their chamber pots! Let's just say that sanitation was not a priority then. (Also apparent was the fact that being a gentleman in merry old Europe was not for die-hard germophobes.)

If people are using things and then throwing things away (creating piles of trash) it won't be too long before the archaeologists show up. Specialized archaeologists (i.e., garbologists)* love plowing their way through society's trash; it is through an analysis of that debris that scientists can get a fairly good picture of the folks who got rid of that debris in the first place.

---

* Garbology is the study of modern refuse and trash and is a bona fide scientific field. If interested, you can concentrate in this field at the institution that pioneered the discipline (in the 1980s), the University of Arizona (which is, as you may recall, this author's alma mater).

Ancient peoples, particularly those who lived along shorelines, relied a great deal on the bounty they obtained from the sea. It therefore stands to reason that if folks are eating lots of sea life, then those same folks are also creating lots of sea life debris, such as bones and shells. It also stands to reason that those bones and shells would be deposited along the shore-lines from which they were obtained—perhaps a pile of debris outside a family kitchen or in a central location on the perimeter of a village, for example. Over the years, decades, or even centuries those piles of sea life garbage might be quite extensive and quite expansive. Similar to the trash accumulations of pack rats, archaeologists refer to these garbage collections as middens (they may also be known as kitchen middens or shell heaps).

Let's take a look at those middens in a little more detail.

The term "midden" means "waste mound" in English. However, it was originally derived from the Danish word *mødding*—a combination of *møg* (muck) and *dynge* (heap). It was the Danes who first began archaeologi-cal studies of these trash heaps in the latter half of the nineteenth century. Today, archaeologists around the world use the term to refer to any accu-mulation of waste products (over time) that might be generated by humans.

> **FAST FACT:** "Midden" is also a word used by farmers in Britain to describe the place where farmyard manure from cows and other animals is collected. Monetary grants are sometimes available from the government so that farmers can construct barriers to protect these middens from rain, thus avoiding runoff and subsequent pollution.

Middens are frequently random collections of domestic waste. That waste may consist of, but not be limited to, animal byproducts, shells, bones, human excrement, pottery shards, stone tools, various artifacts, household goods, specific ecofacts and any other items associated with day-to-day human life. In some cases those collections may be the byprod-ucts of nomadic groups who dump their trash only for a short period of time. Other middens may be long-term accumulations of wastes used by sedentary communities who remained in a specific location for multiple generations.

It's important to keep in mind that all middens were not created equally. Although a midden may be a collection site for the debris of human activity, its specific purpose may vary from place to place. For example, some middens are processing centers—places where food items were prepared for storage or transportation to distant locations. Other middens may be associated with a specific house or dwelling in a village. The debris left behind would be indicative of the foods and other artifacts used by an individual family or family group. Then again, some middens are related to entire villages or towns, a central point in which all the debris produced by all members of that enclave is deposited in a single location. Such piles, as you might imagine, would reveal an incredible amount of information about the dietary habits of a village as well as the products used in their everyday living. Since humans tend to generate trash or garbage on a daily basis, such middens would contain a detailed record of the food consumed as well as the life styles of the inhabitants.*

Shell Middens

Although middens come in all sizes and composition, we are primarily interested in one type of midden—the shell midden. As you might suspect, these sites are large accumulations of invertebrate shells. They may include a specific type of shell, clamshells, for example. Or, they may include a wide variety of shellfish remains (clams, mussels, oysters, whelks), all indicative of a varied seafood diet and, of course, the availability of various types of seafood when a particular area was inhabited.

Shell middens at Grand Bay National Wildlife Refuge in Mississippi.

---

* In 1984, an article in *World Archaeology* described a series of unusual donut-shaped shell middens, consisting of prehistoric pottery, artifacts, and numerous shell fragments. These donut middens were located on hillsides throughout New England. For the longest time, their origin and use was a complete mystery. Then, the author of the article figured out that they were clear evidence of early settlers reusing prehistoric shell deposits as fertilizer for apple orchards. The hole in the middle was where the apple tree stood!

Shell middens can be found throughout the world wherever humans occupied a coastline, a river bank, a lagoon, a bay or harbor, a tidewater flat, a small stream, or wherever some form of shellfish could be found. Many shell middens throughout the world have been dated to the Late Archaic and Late Mesolithic periods (10,000–4,000 years ago). Archaeologically speaking, this was a most interesting time in human history. Although humans were essentially hunter-gatherers during this period of time, they were also beginning to settle down, establishing permanent communities that relied on the ready availability of a broad range of local food resources. Large herds of game, farmable land, and the abundance of shellfish along the coast led to the rise of numerous sites where food was concentrated and plentiful. And, where middens developed.

> **FAST FACT:** The west coast of Canada has numerous shell middens, several of which are more than a mile long. The midden in Namu, British Columbia, is over nine feet deep and spans over 10,000 years of continuous occupation.

Now, let's return, for a minute, to all that trash piling up in your kitchen—specifically, the remains of all the meals you and your family have consumed over the course of, say, the past week or so. Not that I would want to do so, but if I were to forage through your garbage can, I could gather an incredible amount of biological information about you and your family. And just like any professional archaeologist searching through a shell midden left by an ancient people, I could use your garbage to deduce a great deal of data about how you live.

For example, I could use your food remains to gather information on how you traditionally prepare or process your food. Do you eat a lot of fried food or do you prefer to steam or bake much of your food? If your trash primarily consisted of charred meat bones, I'd have a pretty good idea of your food preparation preference. By the same token, if I discovered lots of soggy vegetable matter in your trash, I would know something else about your culinary techniques.

A journey through your trash might also tell me something about the kinds of food (seafood, game, crops) you tend to eat at certain times of the year. Do you rely on vegetables (crops) in the winter or do you depend on game (fresh or stored) as your primary food? I could also determine if, over long periods of time, you tended to focus on one type of food to the exclusion of other types. Were there years in which you and your family relied on animal foods to the exclusion of crops (due to droughts)? Or does your trash reveal a sudden absence of animal remains indicating the possibility of an entire food source being exterminated?

The availability of a primary food source also reveals information about the type of diet a civilization may rely on. For example, if I find lots of bones and animal gristle in your trash, then I'm going to assume that your family is primarily carnivorous. On the other hand, if I locate the remains and residue of varied vegetable matter, I might assume your family is predominantly herbivorous. By the same token, I might see a mix of animal skeletons along with seeds and fruits and gather that your family was omnivorous.

While I'm wandering around through your trash, I can also get a fairly good idea of your general overall health. Let's say I turn up lots of potato chip bags in that tall pile of garbage just outside your kitchen window (not that I'm trying to draw any conclusions about your family's sanitary habits). I know that potato chips have less healthy ingredients than, say, an equivalent amount of venison. That might lead me to conclude that your exclusive reliance on greasy, salty foods would provide you with less nutrition than would a diet that had a balance of several food items. While potato chips might be easier to obtain at the local gas station, a diet that relies exclusively on snack foods might reveal a few dietary deficiencies—something that might show up in the family skeletons over in the village graveyard.

While I'm now firmly stuck in the middle of your trash heap, I might also find some evidence of how you used all those clamshells. Since you're a member of a highly intelligent tribe, you would want to get as much use out of your resources as possible. Perhaps you used clamshells as tools (scrapping, digging, carving). You may have used them for certain types of religious ceremonies (who knows—small clamshell deities positioned on the dashboard of your car might be one of the religious practices of your tribe) or in the construction of some of your village buildings (a dining room constructed of clamshells might be interesting). You could have used them as weapons (sharpened, they could be fastened to the ends of spears). Suffice it to say, how you used the remains of your meals would give me some important data about your lifestyle, your habits and your behaviors, especially over time.

> **FAST FACT:** Early peoples often used everyday objects such as clam-shells to create tools and other specialized implements. These "appliances" were so solid and reliable that many of our modern-day objects are constructed in similar fashion. For example, lightning whelk shells were used like hammers and large clamshells were used for bowls by numerous shoreline civilizations.

For more than a century, archaeologists have recognized shell middens as scientifically valuable remains of prehistoric peoples. Indeed, midden contents are often clear indicators of human cultural history.* Just as important, middens reveal significant clues about how various peoples lead their lives—not just for a short period of time (a week or a month), but over longer periods of time, too.

However, it's time for a word of caution here. (Sorry, it's the old professor in me). We'll call it the "Snickers® warning." Let's say you were walking through the woods and came upon a cache of wrappers for Snickers® candy bars. For our purposes, let's say there were about seventy-five to a hundred discarded Snickers wrappers in a big pile beside a large tree. Being a most rationale person, you would assume that several people were really "pigging out" on Snickers bars and left all the wrappers in a heap in one place. Besides the fact that these people were not environmentally conscious, you might also conclude that "Snickers festivals" were a regular part of their lifestyle or culture.

> **FAST FACT:** The global sales of Snickers candy bars exceeds three billion dollars a year and, until 1990, Snickers bars were sold as "Marathon bars" in the United Kingdom and Ireland.

But, let's say the people who left the wrappers behind didn't actually eat the Snickers bars at all. Instead, they used the bars to build a makeshift dam across a small stream in the woods in order to trap migratory fish.

---

* The oldest shell middens in the world are about 140,000 years old, from the Middle Stone Age of South Africa. One of the most significant sites is Blombos Cave, about 186 miles east of Cape Town. Artifacts unearthed here reveal a higher level of technological advancement, greater ecological niche adaptation, and more sophisticated tool construction than had previously thought possible by early humans.

Or, they used the bars as bait, sling-shotting them into some nearby trees to attract squirrels (which would be subsequently caught, barbequed, and consumed). Or, they used the bars as part of some full-moon pagan ritual involving the ceremonial piercing of selected Snickers bars with extremely sharp needles. ("Take that, you chocolate and nougat demon!") The "eating the Snickers bars and leaving a big pile of wrappers behind" theory seems to be the most logical explanation, but we must consider the fact that eating is not the only explanation for how or why those candy bars were used.

The same thing holds true for clamshells. You would naturally assume that if there were a large pile of empty clamshells next to a village, the people who put those clamshells there consumed the critters that lived inside the shells. That seems very logical. However, we may also want to look at clamshell middens with a more judicial slant. That is to say, throughout history, clams (and their shells) have been used in multiple ways (besides just food). Consequently, it behooves us to examine those middens with an eye, not just on dietary predilections but also on other types of personal or social behavior. While our first thoughts may be to associate clams with food, they can also be associated with other actions or activities. These may include (but not be limited to) varied uses, such as tools, weapons, utensils, building material, jewelry, decorations, religious symbols, and body adornments.

Now you know why some scientists start jumping up and down whenever they come across a new shell midden.

American Shell Middens

Although our focus here is on middens reposing along the shorelines of this country, it's important to remember that clamshell middens are not exclusive to the United States. Almost every seaside country in the world has evidence of shoreline occupation as well as the remains left behind by their shellfish-loving ancestors. As you might imagine, many of those ancient middens are considerably older than those in the United States but certainly no less interesting.

What makes American clamshell middens so fascinating is that they are sprinkled all the way down the east coast of the United States, from Maine down to Florida;* throughout the Gulf Coast from Florida around

---

* Along the New Jersey coast, for example, archaeologists have recorded more than fifty separate shell middens from Sandy Hook in the north down to Cape May in the south.

to Texas; and along the entire west coast from southern California all the way up to, and including, Alaska. Our shorelines are rife with middens. In fact, it would be safe to say that there are literally thousands of them along every shoreline, beachhead, and oceanfront of the United States. (Our ancestors certainly did like their clams.)

> **FAST FACT:** Most shell mounds (on the east coast of the United States, for example) have been dated at 3,000 years old or less. Nevertheless, there are a few that are nearly 5,500 years old.

While I could not even begin to describe all those shell middens (this book would be prohibitively long and prohibitively expensive—even for the most die-hard clam aficionado), below is a very small sampling of selected shell midden sites from throughout the U. S.

Delaware

The close interrelationships between humans and clams are supported by ongoing research at the University of Delaware. Piles of seashells near Cape Henlopen, a spit of land jutting into Delaware Bay near the town of Lewes, aren't just debris washed ashore by historic storms. "[They're] ancient trash heaps from harvests," explained researcher William Chadwick. "The Native Americans would wade into the water and dig up seafood. Then, they would heat rocks in a fire and drop the meat and rocks in a pot full of water to cook."

Combining geology and archaeology, Chadwick's research marks one of the first uses of high-tech Ground Penetrating Radar (GPR) on an archaeological site surrounded by a salt-water marsh. "This technique lets you see what the site looks like before you excavate," he explained. Archaeologists have previously used GPR to examine buried underground foundations, Chadwick said, and geologists use the same equipment to identify subterranean features, such as aquifers.

"Many of the Cape Henlopen shell midden sites lie beneath the present marsh and relict dunes, but this site was identified by the Delaware Department of Natural Resources and Environmental Control because it lies exposed at the surface," Chadwick said. Shards of pottery, pieces of stone tools, and fire-cracked rock were among the diagnostic items found on the surface of the site, which is listed on the National Register of Historic Places.

It is most likely that Native Americans cooked seafood on the site "in a pretty wide time-range, from maybe 1,000 AD until 1,600 AD, long before the time of their first contact with Europeans," Chadwick stated. Many piles of shells were found along the shoreline when settlers arrived, but the numbers have since dwindled. "Geologists in the 1800s suggested farmers use the shell in the middens to lime their fields," he said. "Farmers came down in wagons and picked them up. Now, these sites are protected."

South Carolina

In the ACE basin region of South Carolina (an area divided by the Ashepoo, Combahee, and South Edisto Rivers) multiple shell middens have been identified as belonging to various prehistoric peoples. These people originally inhabited the marshes, wetlands, hardwood forests, and riverine systems of this vast ecosystem, an area encompassing approximately 350,000 acres. Coincidentally, it is also one of the largest undeveloped estuaries along the Atlantic coast.

Dating of the various middens from this region has revealed ancient civilizations from the late Archaic (4000–1000 BC) to the middle Woodland (0–500 AD) periods. Scientific analysis indicates that the peoples who lived here had a varied and eclectic diet. The overwhelming diversity of remains suggests they were the result of food consumption; although there may be some evidence of other uses, including tool making, cooking implements, or hunting paraphernalia. The trash heaps reveal the remains of oysters, clams, periwinkles, whelks, moon snails, razor clams, mussels, cockles, white-tailed deer, minks, raccoons, opossums, black bears, rabbits, turkeys, ducks, several species of turtles (including the diamondback terrapin), blue crabs, and fish (drum, snapper, flounder, catfish, and gar)—certainly a diverse (and quite healthy) diet for prehistoric peoples.

As with many coastal shell middens, the ones in the ACE Basin are subject to the vagaries of weather, climate, erosion, and, unfortunately, human intervention (read: looting). In the centuries since the middens were created, many sites, constantly exposed to the elements, have been washed away or significantly reduced. These valuable historical records are, in many cases, being eradicated faster than they can be researched and documented. As a result, the locations of many sites have become closely guarded secrets.

Texas

Texas was once inhabited by numerous Native American tribes who settled along the Gulf Coast to take advantage of the abundance of natural resources.

As a result, there are numerous shellfish middens ranging from small (three to four individuals consuming a single meal) to very large (millions of shells in mounds more than five hundred feet long, one hundred feet wide, and up to ten feet high). The great majority of the debris in these mounds are of the brackish-water clam *Rangia cuneata*, with most of the remainder comprising shells of the bay oyster, *Crassostrea virginica*. A few other estuarine mussels, clams, and snails also occur, as do both marine and freshwater shellfish species.

Most of the middens in Texas lie along the coast from Corpus Christi Bay northeastward to the Louisiana border. Because shell middens were once Native American campsites, they also contain bones of fish, mammals, reptiles, and other vertebrates along with artifacts such as projectile points, potsherds, and other tools. At present, there is no convincing evidence that the Texas middens were formed through any activity other than subsistence.

The two oldest known shell middens on the Texas coast are located near Galveston Bay and were initially occupied as Native American campsites about 3,500 years ago. Older shell middens are thought to exist, but unfortunately, they are now buried under water and sediment because of shifting tides and frequent storms along the coast.

California

On your next visit to San Francisco, climb into your car and drive east across the Bay Bridge. Shortly after arriving in the East Bay, you'll note that I-80 makes a swift turn northward. Take the first exit, make a right-hand turn onto Powell Street, travel for about a half mile and then take another right turn onto Shellmound Street. You've now arrived in the oft-bypassed city of Emeryville,* a freeway-bracketed town at the junction of two of the busiest thoroughfares in northern California. It's also a town that hosts a most revealing part of California history.

It is here you will discover the Emeryville Shellmound, a historic, cultural, and sacred site established and inhabited by the Ohlone Indians from approximately 500 BC to 1700 AD (long before the rush of cars and the cacophony of urban life intruded into this once pristine environment). The Mound was, at one time, more than sixty feet high with a diameter of more than 350 feet. The Mound dominated the marshy shore and the surrounding landscape, much as the great cathedrals of Europe once towered over the hamlets and burgs scattered across the rural countryside.

---

* Emeryville (not San Francisco) is the western terminus for Amtrak's *California Zephyr*, the train that arrives each day from Chicago.

FAST FACT: Turtle Mound at Canaveral National Seashore (south of New Smyrna Beach, Florida) is the tallest shell midden in the United States. It has been estimated that the two-acre site contains more than 35,000 cubic yards of shells, extends over six hundred feet of the Indian River shoreline, and currently measures over fifty feet in height. In prehistoric times, it was at least seventy-five feet high. Visible for miles offshore, the mound has been used as a navigational landmark since the early days of Spanish exploration.

The view from high atop Turtle Mound.

While this single shell mound was the most dominant one in the area, it wasn't the only one. The Ohlone created a series of nearly four hundred shell mounds ringing the bay. These mounds were, for the most part, composed of abalone, mussels, and clamshells, a staple of Ohlone diet, along with sediment, ash, and rocks. Over time, the mounds grew larger and taller from successive use. They were gargantuan, an enormous accumulation of shells, animal and human remains, ceremonial burial objects, everyday artifacts, and various architectural remains.

> **FAST FACT:** The Emeryville Mound was also distinctive in that it was where the Ohlone buried their dead. Excavations by archaeologists in 1924 reported more than seven hundred burials at the site. These mounds preserved the sacred remains of their ancestors (because the high lime content of the shells). Many of those remains could be traced back for several generations.

The Emeryville Shellmound was occupied for several thousand years. It had been abandoned, however, long before the arrival of the Spanish in California in the 1700s.

Pineland Archaeological Site in Florida

Pine Island is the largest island (seventeen miles long and two miles wide) on the west coast of Florida. Lying twenty miles west of Ft. Myers, it stands in stark contrast to the tony communities of Sanibel and Captiva just to its south. This is a rural island, one with a secluded, quaint, and "out of the way" atmosphere that is as much a part of its geography as its charm.

Residents of Pine Island are proud to say, "Yes, we have almost no beaches," and visitors looking for great expanses of traditional touristy beaches fringed by T-shirt vendors, "Aunt Betty's Homemade Fudge" stands, and kitschy gift shops will be sorely disappointed. Instead, the edges of Pine Island are bordered more by long strands of mangroves than reposing beaches—a factor that draws a different kind of visitor: those eager to step back to a quieter, more rural Florida that existed long before the laying of massive interstate highways and the construction of the state's ubiquitous trademark: mega-sprawling shopping centers.

The island, along with much of the state, originally rose from the receding seas some 24 million years ago at the end of the Oligocene epoch. This was a time when sea levels dropped significantly and much of Florida surfaced from the surrounding waters. However, sea levels continued to fluctuate throughout the termination of the Oligocene and the beginning of the Miocene epoch; as a result, clays and sands became common deposits. In short, islands were created.

Just another day in paradise (or Pine Island).

FAST FACT: One of the predominant animals swimming around Florida during this time was the giant *Caracharodon megalodon* shark. Teeth from this creature, which was over fifty feet long, were more than six inches long. They are still found (though rarely) on Florida's beaches.

While it is not known when humans first set foot on Pine Island, skeletal remains have been unearthed dating back some 6,000 years. Today, Pine Island still offers a distant bucolic atmosphere amid the shore-hugging mangroves, acres of palm, tropical plants and fruit groves, and three aquatic preserves. It is the waters of Pine Island Sound that draw fishermen, bird-watchers and nature lovers of all stripes.

I have come to Pine Island in mid-January, in the midst of the off-season, specifically to visit the Pineland Archaeological Site and view its shell mounds. Cindy Bear, the coordinator of programs and services for the Randell Research Center at the site, has agreed to fill me in on the area's history and discoveries. It is a slightly overcast morning, but considerably warmer and more welcoming than the bone-numbing Pennsylvania winter I left behind the previous day.

Before sitting down with Cindy, I join a group of about twenty other visitors to set out on the Calusa Heritage Trail, a sinewy footpath that meanders nearly a mile through the shell mounds, canals, and other features of the Pineland Archaeological Site, a massive mound complex spreading over more than two hundred acres. The trail ribbons its way through this ancient site offering numerous vistas, historical artifacts, and detailed information regarding the Calusa Indian people who inhabited the landscape for nearly 1,500 years.

Our leader is Jim Fors, a long-time docent with the research center. With a classic safari hat, an equally classic safari vest, and a jaunty (safari) attitude, Jim is the epitome of a well-versed tour guide. His information is rich, and his delivery is enthusiastic; for the next hour and a half he keeps the group moving and he keeps us well informed.

After a thorough introduction on the Calusa Indians, Jim offers a time-line of the site. Like a practiced kindergarten teacher, he competently herds the diverse group past historical venues and natural history views while peppering his presentation with significant facts and mesmerizing data.

One of the first vistas we come upon is Brown's Mound Complex, a towering shell mound that began to accumulate about 1,500 years ago. Jim informs us that the mound was once much larger, extending well over thirty feet high. At its summit may have been the residence of the town chief and other nobles. A sign posted near the mound offers additional information: "Height often indicates power, authority, or wealth. In our own society, we recognize the prestige of 'the house on the hill' and big businesses imply success with tall buildings. Pineland's residents probably also built houses on high places for practical reasons: protection from storm surges, biting insects, and human enemies."

FAST FACT: According to the American Mosquito Control Association, in general, mosquitoes that bite humans prefer to fly at heights of less than twenty-five feet.* Thus, a house situated on a mound thirty feet high would mitigate much (but not all) of the mosquito problem.

---

* For those inclined to consume beverages brewed with hops, barley, and malt during hot summer months, the American Mosquito Control Association also notes, "People drinking beer have been shown to be more attractive to mosquitoes." Don't say you haven't been warned!

The group climbs up an adjacent set of wooden steps to get to the top of Brown's Mound. Jim explains some of the natural history of the area and then takes us along a short ridge to a platform at the summit of the mound. From here we can see for many miles in all directions. Jim points out the obvious benefits of such a view—watching for any approaching enemies, keeping an eye out for possible storm surges from offshore, and searching for any herds of animals that might be wandering past the town (and, hopefully, into the town's cooking pots).

Jim also points out a sign on the edge of the small platform quoting Frank Hamilton Cushing, one of the first archaeologists to visit the site in 1895:

> Climbing to the summit of the heights of this mound, I was aston-
> ished beyond measure with the extent of the works which became
> visible there from. To the southwestward, as far as the eye could
> reach, to the southward stood a succession of these great shell
> heights with their intermediate water quartz, graded ways and
> canals, with their shelly surrounding platforms and terraces (hand)
> to the southeastward, yet another series of the gigantic heights.

We meander around the site for a few minutes more, and Jim leads us back down the steps and past an adjacent, though considerably smaller, shell mound. It was at this mound that archaeologists discovered evidence of a 700-year-old house. There are the remnants of posts where walls once stood, but it is the floor that may be the most interesting aspect of this particular site. Mixed into the floor, archaeologists have discovered several exotic materials (such as galena from Missouri). These materials may have been used for body paint, suggesting that the house might have been a place high-status people used to prepare for rituals. Additional excavations are planned to learn more about this shell-mound dwelling.

Jim gently herds the group along the trail past several other sights, including remnants of an extensive canal system created by the Calusas. The canals were each approximately thirty feet wide and six feet deep and were used to transport cargo across the length and breadth of the island. Their extreme width made it possible for two cargo canoes to pass each other side by side. However, what I found most impressive was that this extensive and elaborate system of navigable canals was dug out of the Florida landscape entirely with, you guessed it, clamshells!

As Jim shepherds everyone along the trail, sharing additional infor-mation about the Calusas, their distinctive structures, and their complex society, I exit out the back end of the group and trek over to the Research Center for my scheduled appointment with Cindy.

When Europeans first arrived on Florida's shores in the 1500s, the Calusa Indians were the most powerful people in all of South Florida. Inhabiting an extensive region—now today's Lee, Charlotte, and Collier Counties— the Calusa were a prosperous, powerful and artistic Native American people who built towns, engineered canals, and developed a complex society that was, in many ways, more sophisticated than most in the known world.

From Pineland, the Calusa exerted influences in several different directions. They often received tribute or payment of goods or services from as far away as Lake Okeechobee, present day Miami, and the Florida Keys. In the 1560s the Calusa domain included at least fifty towns and, by 1612, the Spanish governor, Juan Fernandez de Olivera, reported that the Calusa leader controlled more than sixty towns.

Central to the Calusa domain were the waters of Charlotte Harbor, Pine Island Sound, and Estero Bay. The centerpiece was Pineland, a major, centrally located Calusa village for more than 1,500 years. Enormous shell mounds still overlook the waters of Pine Island Sound. The remains of fifteen centuries of Native American life are evident everywhere. Remnants of the ancient canal system reach across Pine Island; in addition, two shore-line mound complexes, and several burial mounds stand as testament to the influential Calusa society. Pineland was one of the three largest known Calusa towns. Only Big Mound Key in Charlotte County and Mound Key near Fort Myers beach are of comparable size.

FAST FACT: When Spaniards first arrived in south Florida, about five hundred years ago, Pineland was a prominent Calusa town called Tampa. The name was still on maps as late as 1683. However, in the 1700s, mapmakers incorrectly (or myopically) moved the name eighty miles north (most likely due to geographic and cartological similarities between Pine Island Sound/Charlotte Harbor and Tampa Bay). Today the name is associated with the modern city of Tampa.

The Calusa were not farmers but instead prospered from the bounty of the subtropical coastal environment. Native plants and animals characteristic of coastal hammocks, pinelands, wetlands, and shell mounds were in abundance. For many centuries, the Calusa sustained thousands of people from the fish and shellfish found in these rich estuaries. Lacking local stone or metal, they developed tools and ornaments from shell and bone. And, their painted and sculpted wooden artwork is among the most renowned in North America.

The Pineland Archaeological Site is listed in the National Register of Historic Places. It includes the Calusa Heritage Trail and the Randell Research Center, where archaeologists have been conducting research since 1988.

"Shell mounds—midden mounds—record the past. They record environmental conditions of the past. They record the technology of the people of the past and change in time. They give us a link, in a very material way, to what our environment has been, how people have lived in that environment and changed it, and how it's changed them."

I am chatting with Cindy Bear in the spacious classroom of the Randell Research Center at the Pineland Archaeological Site. An energetic and passionate woman, Cindy punctuates her conversation with large exclamation points and a plethora of insights, solid research, and an enthusiastic sense of mission.

After obtaining a degree in wildlife ecology at the University of Florida, Cindy found herself working in environmental education for a local school district. Later, a master's degree in science education and membership on the advisory board for the Randell Research Center helped her create a collaborative partnership among archaeologists, educators, and ecotourism operators. It was here that she learned the interdisciplinary nature of archaeology and how to transmit that information to children as well as to the general public.

When her position with the school district was restructured and the opportunity presented itself for her to pursue her greatest passion—outdoor education—serendipity intervened and a new position, coordinator of programs and services at the Randell Research Center, opened up. "I needed to get back to my true roots," Cindy tells me. "I needed to be where I could be most effective for an organization." And, since 2010, she has brought that determination and passion for teaching to the thousands of visitors who come to the Pineland Archaeological Site each year.

With my notes on the Calusa Indians before me, I ask Cindy to share what archaeologists have learned about these people through their studies of midden mounds, specifically the midden mounds at Pineland. She tells me that frequently what is conveyed in the media is often conjecture or inference. There might also be some ethnology or historical documentation. Or, the reporting might be a combination of all those factors. She is quick to remind me that, for the sake of accuracy, "it is important to distinguish that which we see, or measure, or have verifiable evidence of, from that which may be speculation."

As an example, she refers to Brown's Mound. Because it was such a massive mound, it was originally speculated that it must have been for the person with the highest status and greatest power. However, as archaeologists began looking for evidence in the midden mound, they discovered no solid evidence to support that claim. So that point is still speculative. However, what they did find was material that had been thrown away in one place and later transported to Brown's Mound. Radiocarbon dating showed that older material was on top of younger material—an "upside down" effect, if you will. In short, the mound was not built sequentially, but rather haphazardly.

Further, Cindy tells me that there is clear evidence that the mound also grew as the result of the discard of materials over time, and those materials included leftover mollusk shells, fish bones, and pottery shards. This is the material evidence (as opposed to conjecture) that archaeologists need to confirm hypotheses. Cindy refers to these as the "signatures" of environmental conditions. As examples, she offers the following: "For instance, the shells of quahog clams contain isotopes whose ratios correspond to water temperature at the time the animal was alive. The size of the vertebrae of pinfish bones in the middens relates to whether the animal died in spring, summer, fall, or winter and hence this is evidence to confirm that people lived at the site year round rather than the site being a place for special events only."

What also ends up being recorded, throughout this midden mound, are the responses of a group of people to changes in the environment. Cindy emphasizes that that is critical for all of us today. She points to the fact that archaeologists can see shifts in patterns of technologies to catch food, patterns of what people are able to gather and use. Evidence found of other technological developments, such as tool making, can also lead to inferences or conclusions about how and why these people achieved power and strength and why a community of fisher people became a powerful state with a highly structured class system.

For Cindy, it's part of her educational process—to understand what can be learned from evidence in the ground and what can be inferred based on additional evidence from other sources, including historic documents. She tells me that there has been a tendency in the past for some people to refer to the mounds as "temple mounds" or give them other such designations. Yet it appears (by evidence) that these were primarily domicile mounds, places where people lived. She adds, "And what a great place to live."

> **FAST FACT:** Temple mounds have been found throughout Mesoamerica. These artificially created mounds of earth usually had a shrine or temple on top. In some cases, they were used as burial plots.

My own research had revealed a plethora of discoveries awaiting archaeologists bent on examining Florida's midden mounds. I was curious, therefore, to learn of the most significant discoveries made in the mounds at Pineland.

"Here, the most significant discoveries are that archaeologists can discern and detect environmental and cultural changes on a much shorter time scale than what's expected," Cindy tells me. "They can detect change in increments of fifty to one hundred years. That is really significant and important when you're thinking about human life times. Often it's been assumed that change could only be detected on a much longer time scale and/or that environmental conditions were stable. Archaeologists have documented here that, in fact, change in many forms can happen abruptly and that can make differences culturally as people live on a landscape." As I listen to Cindy, I think about the abrupt changes that have been made in my lifetime—technological changes—and how those changes have altered (and multiplied) the ways in which I can communicate with others, for example.*

Cindy also relates how, when archaeologists began working here in 1989–1990, it really wasn't clear whether people lived here year round or not. Nor was it clear what their strategies were for fishing. Did they go out in boats and use hooks and line or did they stay near shore and use nets? It has been through an examination of shell middens that the answers to those questions have revealed themselves. It also underscores the significance of middens as "archaeological treasure chests" of tangible data (as opposed to rampant speculation) validating behaviors that were practiced many centuries ago—long before they were recorded with pen and ink.

Cindy's pace quickens and her natural enthusiasm ramps up as she tells me about the speculation about the Calusa and their lack of agricultural sophistication. However, through archaeological research, "we've been able to refine our statements about the Calusa and their use of plants."

---

* And the multiple ways others can communicate with me: offering "exclusive" deals in southwestern real estate, "once-in-a-lifetime" timeshare investments, and special opportunities in retirement living for my "golden years."

As Cindy tells me, it was once believed that the Calusa didn't grow or manipulate any plants. However, as it turns out, archaeologists have discovered 2,000-year-old papaya seeds at Pineland. Botanical examination of these seeds has revealed unique "signatures" indicating a purposeful manipulation of the plants. In short, the Calusa engaged in some form of gardening. Without that archaeological evidence, the gardening hypothesis would have been just what it was—an educated guess.

> **FAST FACT:** Research by the Florida Museum of Natural History has shown that the Calusa raised no corn, beans, or manioc—staple crops considered essential by many Native American peoples.

Our time is drawing to a close, and so I ask Cindy to tell me what she has found to be the most fascinating aspect of her time at the center.

"Personally, the one thing I have found fascinating here, that I didn't expect, was how interdisciplinary the science of archaeology is." She tells me that archaeology is actually a combination of the work of many scientists including botanists, zooarchaeologists, naturalists, historians, geologists, language experts, and biologists. "It's just an incredible discipline that spans so many sciences and that brings it all together."

As I tidy up my notes and switch off my recorder, Cindy reminds me of an old archaeological maxim: "It's not what we find, it's what we find out—that's what's important!" As I stroll down the shelly path to the entrance gate, I am convinced that the Randell Research Center will continue that scientific tradition as a place that will keep making discoveries, preserving the past, and teaching the public about the significance and value of shell middens.

In other words, the debris of ancient peoples, just like the contents of your trashcan, is always revealing, always insightful.

C h a p t e r 3

Bivalve Legends

Legend: A lie that has attained the dignity of age.

—H. L. Mencken

W HEN I WAS A YOUNGSTER, I SAW A SWIMMER attacked and unmercifully trapped in the jaws of a giant clam. Unbelievable!

In the 1950s, giant creatures ruled the movies. This was an age when atomic monsters were created, weird prehistoric behemoths were awakened from extended hibernations, horrible mutant life forms would take over the bodies of unwilling farmers in Iowa, and prehistoric and amazingly oversized reptiles would knock over a couple of office buildings in the great metropolis shortly after emerging from a multi-million-year slumber.

The cinematic "rule" of the era was the larger the creature, the better the movie. Even in those days, size made a difference.

Although it was released somewhat earlier than the larger-than-an-aircraft-carrier-giant-monster-with-a-really-bad-attitude-themed films of the fifties, *16 Fathoms Deep* (1948) provided me with additional verification that whopping creatures or humongous critters should be avoided at every possible opportunity. The film profiles a former Navy frogman (Lloyd Bridges of *Sea Hunt* fame) who takes a job as a sponge fisherman in Tarpon Springs, Florida, on the Gulf of Mexico. Signing on with a new skipper, our hero discovers that the job is considerably more hazardous than expected, due to the dirty tactics and underhanded dealing perpetrated by a nefarious sponge dealer (played by Lon Chaney, Jr.).

Near the end of the movie the narrator (Bridges) begins to get a little more excited and the background music starts to get a little faster. It's now time for the climactic scene, the one where a young sponge diver races underwater to save his father who has been trapped in several entangling lines. Unknowingly, the young diver swims just a little too close to an oversized bivalve (identified for any budding marine biologists in the audience as a four-foot-long, five-hundred-pound, *Tridacna*). As is only possible in films with ever-quickening violins and an overly frantic narrator, the giant clam grabs the defenseless diver's dangling foot in its lightning-quick shells (jaws?).

The end is inevitable! The diver struggles, the clam clamps shut, and the background music reaches a full orchestral crescendo. Then (you guessed it), the young and still struggling man is slowly and methodically drowned by the enormous (and what is now a carnivorous) bivalve. The fact that giant clams don't live in the Gulf of Mexico or consume unsuspecting sponge divers did not stop the screenwriters from crafting this piece of cinematic lore for the sake of an exciting story. Normally quite placid, the once languid *Tridacna* has now been saddled with a most ominous moniker: Killer.

A Pacific Legend

Apparently giant clams don't exist just in the minds of Hollywood producers; they have been part of the culture and traditions of many peoples around the world for generations. In Chinese mythology, for example, the word *shen* is used to mean a shellfish, or mollusk, identified as an oyster, mussel, or giant clam. Literally translated it means a "large shellfish" or a "large clam," which is also viewed as a shape-shifting dragon or sea monster—one believed to create mirages.

Further away, about five hundred miles east of the Philippines, is the island nation of Palau (officially The Republic of Palau). Initially settled more than 4,000 years ago, this collection of more than two hundred islands currently has a population of approximately twenty-one thousand people of mixed Austronesian, Melanesian, and Micronesian descent.

FAST FACT: In 1981, Palau voted for the world's first nuclear-free constitution. This constitution banned the use, storage, and disposal of nuclear, toxic chemical, gas, and biological weapons.

Part of the culture and history of Palau rests on its legends that mate the ever-constant sea with tales of creation and myth. One of those stories features a giant clam, an oceanic creature revered by ancient Palauans. The following is a retelling of a favored creation myth.

In the long ago time, in the time before the people and the animals, even long before the time of land, all that could be seen was the vast and empty sea, stretching everywhere beyond the horizon. And the spirits that traveled around and in and over the vast and empty ocean felt great loneliness.

Ucheleanged, the greatest of all the gods, felt the world's yearning. "It is time," he said. "Let us let life begin and all the emptiness be filled." And Ucheleanged turned to point to the very darkest part of the vast sea.

It was then, from the ocean's great depths, that a volcanic mountain rose skyward from the waves, but this mountain was not barren. On the very top, where the spirits of the seas and the spirits of the heavens touched, there, at that very place, sat a giant clam. It was silent. It did not move. It never spoke. But it was growing . . . growing . . . growing.

The giant clam continued to grow and grow, its thick white shell spreading strong and far and its soft, sweet middle bulging large and thick and pregnant with new life. Soon the giant clam was filled and stretched to fullness, ready to pour forth.

But the clam could not give birth.

Ucheleanged sent a mighty storm upon the mountain. Giant waves crashed against the thick shell of the clam. The thunder roared, shaking the sea with dark, dark anger. The vast ocean frothed green and gray, the surf tumbling and tossing, trying to dislodge the stubborn clam. Finally, from the deepest trough of the sea, a mighty tidal current ripped upward, tearing at the fullness of the clam, splitting open its inner body and spewing into the sea all the many forms of life inside. First to emerge were the smallest shrimp and oyster, then crab and lobster, followed by animals that swam and crawled emerged, and then all the birds that flew, a crowing rooster, soaring bats, creeping lizards and gnawing rats, until, finally perhaps, eons later, the pig, cat, and dog.

All the sea creatures gave birth to many, and soon the vast and sweeping ocean was teeming with life. It was after these events that the giant child, Uab, "gave birth to" and became the islands of Palau, each island and village retaining a designated political and social status depending on from which part of Uab's body the land derived.

Giant Clams

Just like all those oversized science fiction monsters of my youth, giant clams have long been wrapped in an aura of mystery and intrigue. The fact that these creatures are considerably larger than your "average" critter may propagate stories with just a hint of believability or a tiny sliver of truth. You see, humans are natural storytellers; we love to take something about which we know very little and spice it up a bit, imbue it with a little fabrication, a moderate fib, or an outright distortion of the facts (sounds like your typical congressional candidate). Although we may not totally understand the real science behind an animal's anatomy, physiology, or behavior we can certainly make the creature a little more intriguing than it has a right to be. And, so we create stories; we create legends.

One such legend surrounds the discovery of the Pearl of Lao Tzu, the world's largest pearl. This particular pearl is also known as a "*Tridacna* pearl" or "giant clam pearl."* The pearl was obtained by one Wilburn Cobb from a Philippine tribal chief in 1939 as "payment" for saving the life of his son, who was dying from malaria. In describing his acquisition, Cobb crafted the tale of a Dyak diver who drowned when the giant clam from which the pearl was obtained closed its shell on the tribesman's arm. Even the prestigious *National Geographic* ("Giant Clam—*Tridacna gigas*") suggests on its website this is not the only legend of South Pacific origin to portray these bottom-dwelling behemoths lying in wait to trap unsuspecting swimmers or swallow divers whole.

In its May 1924 issue, *Popular Mechanics* magazine displayed a murky picture of one apparently evil-looking, enormous (and, what the editors hoped, carnivorous) bivalve and wrote: "Divers who often step into the open lips of such monsters are frequently held with such force that they cannot release themselves and drown. The shells . . . serve as gigantic traps." During World War II, the official *U.S. Navy Diving Manual* gave explicit instructions on how to free oneself from the potentially lethal grasp of this heaviest and most massive of shellfish. The trick was to sever the abductor muscles used to open and close the shell. Of course, this assumes that one had a very long, sharp knife handy and was capable of reaching around the enormous four-foot shells to cut said abductor muscle sufficiently, and in time, to be released from the monster's embrace. Even the author of a 1986 book, *Dangerous Australians: The Complete Guide to Australia's Deadliest Creatures*, perpetuated this myth

* The pearl was so named because it measured 9.45 inches in diameter and weighed in at 14.1 pounds.

when he stated, "Popular legends describe giant clams trapping men with an incredibly powerful vise-like grip and eventually drowning them."

As you know, sometimes when a story gets started it's difficult to stop. It begins to assume a life of its own, regardless of the facts and actual science. Because we may have limited exposure to a far-flung and somewhat obscure species, imagined facts are often inserted into those little gaps in our knowledge base (e.g., if a creature is gigantic, or considerably more monstrous than other, more familiar creatures, then it stands to reason that it will be a threat to humans). Then, if the creature inhabits a part of the world about which we have little or inadequate information, then said creature assumes an identity somewhat different from the norm. Now, we have a legend.

Inevitably, those legends are spread by storytellers across geographical distances and real time. Inadvertently, and unconsciously, the stories may be verified by their appearance in local or regional publications or other forms of media (often because fact checking is not a priority). Over time, myth becomes reality and real science takes a back seat. These stories slowly become a part of the culture, creating both melodrama (for movie producers) and myth (for the general public). As a result, we have fabricated tales about the tendency of giant clams to devour, drown, or dismember unsuspecting human aquanauts who insist on placing key body parts inside seemingly innocent mollusks . . . and suffering the inevitable and quite horrible consequences. Once again, an oversized beast (e.g., octopus, ant, tarantula, creature from another planet, giant clam) becomes an unwitting, if unintentional, human-killing machine.

Although the giant clam has been frequently identified with monikers such as "Killer Clam" and "Man-eating Clam," such a carnivorous reputation is entirely undeserved and absolutely false. In fact, let's set the record absolutely straight here: there is no recorded or verified account of a human death at the hands (or shells) of a giant clam. Ever! Yes, their shells can close just like the shells of the quahogs or Manila clams that we tend to plop into large vats of steaming water. And, it is feasible that those giant clamshells might (emphasize the word "might") be capable of gripping one or more human appendages. But, the fact remains that it has never happened simply because it would take a very long period of time for a giant clam to get its shells from *open* to *closed*, sufficient time, I suspect, for you to redecorate your living room, alphabetize your kitchen pantry, and color code all your socks. By that time, we would hope that any curious swimmers would have had the good sense to remove said appendages from any clam lips and depart the area.

> **FAST FACT:** Giant clams are the largest mollusks on the planet.

But, perhaps, we're getting ahead of ourselves. Let's take an inside look into the giant clam (from a safe distance, to be sure).

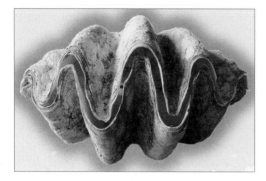

A *Tridacna gigas* specimen from the Philippines measuring 55 centimeters (21.7 inches).

The giant clam (*Tridacna gigas*) lives up to its moniker by weighing in at more than 440 pounds and measuring nearly four feet across (as an adult). By comparison, a fully-grown giant clam would be equivalent to the weight of a pygmy hippopotamus and the length of an adult western diamondback rattlesnake. To scientists, the giant clam is an oversized bivalve, one with a lifespan of up to one hundred years or more. *T. gigas* is typically found in shallow coral reefs of the South Pacific and Indian Oceans (never, I might add, near Tarpon Springs, Florida). They spend their lives in flat coral sand or broken coral at depths of up to sixty-six feet.

The reproduction of giant clams is clearly distinctive in the animal world. You see, giant clams are hermaphroditic. That is to say, they are capable of producing both eggs and sperm. Now, before you get ahead of me, I do want to point out that self-fertilization (or, how shall I put this politely . . . sex with oneself) is not possible. Fertilization takes place only when the eggs of one clam come into contact with the sperm of another clam. I realize that process doesn't have all the charm and magic of what we typically know as sexual reproduction, but it has proven to be quite effective for these marine leviathans. Unlike the twenty-somethings at your local singles bar on a Saturday night, they don't have to spend needless time searching for a compatible mate—everyone around them is compatible. Cool!

Since giant clams are firmly fixed onto the coral reef or other substrate, they cannot move. As a result, they have adapted a process that enables their

sperm and eggs to get together in a very efficient manner. Known as broad-cast spawning, giant clams cast all their sperm and all their eggs into the water during a well-coordinated yet brief window of time. This action is triggered by the clams' release of a substance known as Spawning Induced Substance (SIS)—what you and I might refer to as the ultimate sexual aid. SIS stimulates the simultaneous and synchronized release of all those sex cells, thus ensuring lots of fertilization (and, of course, lots of potential relatives).

After fertilization, the eggs of the giant clam float in the ocean for about twelve hours until a larva eventually hatches. It is at this point that the soon-to-be giant clam measures a mere 0.0063 inches in length. (Several giant clam larvae could actually fit inside this letter "O"). Soon after hatching, the larva begins to produce the chalk shell inside which it will live for the remainder of its life.

At about one week old, a clam larva settles on the reef; however, it may move around during its first few weeks. Eventually the front abductor muscle (the muscle clams use to close their shells together) disappears and the rear muscle moves into the clam's center. It is at this point in its life that the clam becomes sessile; that is, it becomes firmly attached to a final resting place, a location that will be its home for the next several decades or so. From this point on it will grow at a rate of approximately four-and-a-half inches per year until it reaches maturity at about ten years old.*

For most of their lives, giant clams hold fast to their respective coral

> **FAST FACT:** The giant clam has small eyes that can only detect shadows. Thousands of these pinhole eyes lie along the mantle where they function as an early-warning system for predators.

reefs and large underwater rocks. They reach their prodigious size by constantly sucking in water and filtering out single-celled dinoflagellate algae (*zooxanthellae*). The filtered algae take up residence inside the clam. Then, during the daylight hours, the clam opens its shell and extends its mantle tissue so that the algae receive the sunlight they need to photosynthesize. In return, the clam gets to consume sugars and proteins produced by those

---

* By comparison, the physical growth of human children is quite inconsistent. The growth of most babies is characterized by sudden growth spurts, frequently followed by periods of slow growth. On average, however, the height of children increases by two to three inches per year until about seven years of age.

same algae. This is a classic symbiotic exchange: the giant clam offers the algae a safe place to live and the algae obtain the sunlight they need for photosynthesis. You could say it's sort of an "I'll scratch your back, if you'll scratch my back" type of relationship.

> **FAST FACT:** Giants clams have several predators, including sea stars, snails, fish, and eels—all of which may nibble on exposed parts of the giant clam.

Interestingly, giant clams come in a variety of sizes. They may range from the *Tridacna crocea* (or crocus clam)—the smallest species of giant clam (with an average shell length of six inches)—all the way up to the aforementioned *Tridacna gigas* (shell length up to four feet). In fact, the largest known specimen of the giant clam was discovered on the northwest coast of Sumatra in 1817. This specific creature measured approximately fifty-three inches across and had a live weight of about 550 pounds. Interestingly, the shells of this particular specimen are now on display at the Ulster Museum of Northern Ireland in Belfast, where they were donated in 1830. Another unusually large clam was found in 1956 off the Japanese island of Ishigaki. This specimen's length was about forty-five inches, and its overall estimated live weight (it was not scientifically examined until 1984) was 750 pounds.

You might think that extra-large creatures would not be subject to the vagaries of Mother Nature as would, say, mice or insects or tiny fish. Not so! You see, the giant clam, in spite of its prodigious heft and overwhelming presence in various tropical neighborhoods, has been officially classified as "Vulnerable" on the International Union for Conservation of Nature and Natural Resources (IUCN) Red List.* A "Vulnerable" species is one "considered to be facing a high risk of extinction in the wild." Worldwide populations of giant clams are on the decline due primarily to extensive exploitation by fishing vessels that seek to harvest these creatures for their meat or to supply the aquarium trade. Primarily it is the large adults that

---

* The ICUN rates each of the world's plant and animal species on a sliding scale ("The Red List"). There are seven categories on the list: Least Concern (e.g., Greek fir), Near Threatened (e.g., large-mouthed frog), Vulnerable (e.g., bigeye thresher shark), Endangered (e.g., wild almond), Critically Endangered (e.g., knifetooth sawfish), Extinct in the Wild (e.g., Polynesian tree snail), and Extinct (e.g., dodo bird).

are killed, since they are the most profitable . . . and, unfortunately, most edible.

In many countries, including Japan, France, Southeast Asia, and several South Pacific nations, the giant clam is considered a true delicacy. As a result, the meat from the clam's muscles are frequently served as an ingredient in various recipes in numerous Asian restaurants. Besides its appeal as a culinary treat, giant clamshells are sold as decorative items on the black market. These practices, long established in many countries, have raised a genuine concern among conservationists worldwide. Scientists and marine biologists fear that human (over)exploitation of this species may have serious and irreversible consequences for its ultimate survival in the wild.

FAST FACT: Giant clams evolved more than 65 million years ago (about the time dinosaurs died out at the end of the Cretaceous Period).

Despite its "reputation," the giant clam is quite docile and sedate. Even though it is one of the largest organisms on the planet, its gargantuan size is more to be admired than feared. While often cloaked in both myth and mystery, it remains as one of the most incredible animals on the planet. Although it may appear to be the quintessential gourmand's delight ("The Ultimate Steamer") or a potential human-devouring character in a late-night flick on the Syfy channel, its size belies its true character and demeanor.

The Raven and the Clam

For First Nation* and Native American people in the Pacific Northwest, the raven is a potent symbol and holds a prominent role in native mythologies. For many indigenous peoples the mythological Raven takes two forms: one is the creator of the world; the other is a trickster, who fools or plays tricks on others to get his way. According to at least one legend, Raven was responsible for bringing the sun, moon, and stars into being, which he accomplished by tricking Seagull, who had been entrusted with preserving the box holding the light of the world.

The Tsimshian are one of the indigenous peoples of the Pacific Northwest, predominantly located in British Columbia and Alaska. Like many people of this area, the Tsimshian lived, and continue to live, primarily along the

---

* In Canada, native peoples are known as "First Nations" people. In the United States we use the term "Native Americans."

coast. As a result they are highly dependent upon the ocean for much of their survival and livelihood. While salmon fishing consumed much of their time and effort, they also fished for other types of seafood in the rich and abundant waters.

FAST FACT: The Tsimshian originally founded the settlement of Kitkatla in British Columbia. Kitkatla is one of the longest continually inhabited communities (dating back 10,000 years) in all of North America.

Tsimshian culture is filled with stories and myths about the animals with which they live. Most of these stories are oral tales—typically stories about animals in human guise or with human emotions. As is the case with many other cultures, these stories often serve to explain scientific principles or conditions in creative and imaginative ways. For example, Raven often assumed a prominent role in many Tsimshian stories, since he could either be good or evil; but more than likely he could assume his inevitable trickster role—thus making any resulting story both informative as well as entertaining. Many of these Raven stories have survived to the present day.

The following Tsimshian story, adapted from Michael J. Caduto and Joseph Bruchac, features clams—if not in a starring position, then at least in a supporting actor role. You will quickly notice how the clams help Raven achieve his ultimate goal.

Raven flew over the waters until he reached the mainland and the wigwam of the old, old woman who holds the tide lines in her hands. At that time the tide would remain high for many days at a time, so that the people could get no clams or other seafood. It happened that Raven was very hungry for clams, but he entered the hut and sat down, saying pleasantly:

"Good day, Grandmother. There is fine digging today. I have just had all the clams I could eat!"

"Nonsense," exclaimed the old woman. What are you talking about, Raven? You know very well that the clams are all covered with water."

"Yes, but I've had all the clams I want," he insisted.

"That isn't so," she declared.

Upon her saying this he rudely pushed her backward, until she fell down and her mouth and eyes were filled with dust. Of course,

she was forced to let go the tide lines, so that the tide ran quickly out, and the beach was covered with fine, fat clams and other shellfish. Raven did not come back to the hut until he had eaten as many as he possibly could.

"My eyes are blinded with dust," mourned the old woman. "Will you not give me back my sight?"

"I will, if you promise to slacken the tide lines twice a day," he replied.

So she said that she would, and from that time to this the tides have run in and out twice a day.

From Ancient China

According to Chinese folk legend, Guanyin is a mythical woman often known as the Goddess of Mercy, who offers great mercy and compassion to distressed or needy individuals. In the beginning, Guanyin was a male and he dwelled on an Indian Mountain. But when he came to China, he gradually became a female and was called "Goddess Guanyin." In Buddhist scriptures, Guanyin is said to have vast magic powers and is capable of saving people by listening to their voices and liberating them from suffering.

> **FAST FACT:** The images of Guanyin are diversified and varied. She has been often depicted as a beautiful lady, a thousand-armed Guanyin, and a children-sending Guanyin. Another frequent image of Guanyin is one in which she is holding a lotus blossom or a willow twig.

As you might expect, there are many stories of Guanyin. One familiar story about her (827–840 AD) takes place in the capital of the Tang Dynasty: Chang'an (known as Xi'an* today). Chang'an was located in the central portion of China, approximately 679 miles southwest of Beijing. The ruling emperor (Emperor Wenzong)at the time had a particular fondness for clams, frequently ordering the bivalves for three of his five daily meals. He did this, in spite of the fact that the royal palace was a considerable distance from the sea (840 miles) and that it took scores of individuals working together to ensure the safe and timely arrival of the clams.

---

* Xi'an is the capital of Shaanxi province, and a subprovincial city in the People's Republic of China. One of the oldest cities in China, it has more than 3,100 years of history.

Guanyin.

Very early in the morning, clams would be gathered by the fishermen of Zhejiang and then packed by porters in cold seaweed, wet sand, and ice. They were loaded on relay mounts that sped along the Imperial Highway to Chang'an. Obviously, this entailed a great deal of manual labor (and coordination) just to please the emperor's dining predilections.

Day after day, hundreds of laborers toiled to ensure the emperor received his clams—many of them living in constant misery and poverty. Then, an incident shocked the entire palace. One day, the royal chef noticed an unusual clam of grand size in the daily delivery. The clam was enormous, twenty times the usual size. It was surely an imperial clam meant for the imperial palate. However, as the clamshell opener stepped up to pry the shell apart, he discovered the shell sealed like iron.

Emperor Wenzong heard about this and commanded the opener to let him take a closer look. As the emperor approached, as if by a silent signal, the clam began to slowly open. The emperor was astonished at what he saw. There, standing inside, was a finely detailed, miniature, and astonishingly sweet statue of the Goddess of Mercy, the Bodhisattva Guanyin, exquisitely carved. What surprised the emperor even more was what Guanyin said by her lovely expression: "The laborers have to make great efforts for your own pleasure; you both harass the people and waste money."

It was at this moment that the emperor realized that Guanyin, this Buddhist goddess, who hears even the smallest call for mercy from even the tiniest voice in the empire, had taken pity on the boat men, the fisher folk, the portage men, and the relay riders. She even took pity on the royal cooks—all those who served the emperor's royal taste and royal whim. The emperor suddenly realized the reason Guanyin showed up in the grand clam was to warn him. It was clearly evident that she would always be there to watch over mankind, particularly those who needed her presence most. Once again, a legend prevails.

Aphrodisiacs—The Persistent Legend

Will eating clams improve my love life?

Maybe yes. Maybe no. (You'll note how definitive certain authors can get sometimes.)

Ever since the dawn of civilization, humans have been looking for remedies, solutions, and magic potions that will increase their passion, inflame their desires, and satisfy their basic urges (sounds like the plot line for an afternoon soap opera). In short, we want better sex . . . or more sex. Can nature help us out?

Named after Aphrodite, the Greek goddess of love, beauty, and fertility, aphrodisiacs are substances that supposedly elicit sexual desire and arousal, enhance sex drive and sexual "performance," and extend sexual energy. For centuries, humans believed that shellfish were natural aphrodisiacs. For example, your basic Roman orgy was an opportunity to engage in indiscriminate sex, but it was also an opportunity to serve lots of shellfish—the belief being that consumption of certain bottom-feeding invertebrates would raise one's overall libido to new levels (certainly an advantage if one was planning to drop in on several orgies in the same evening).

The connection between shellfish and sexual desire (or prowess) was further accelerated by legendary lothario Casanova, who reportedly consumed sixty shellfish every morning for breakfast, and then spent the rest of the day in a tub entertaining one or more young ladies in a variety of passionate encounters. His ability to sustain that heavy dose of lovemaking was attributed, in large measure, to his morning diet (which, you must admit, sure beats corn flakes).

Casanova, the original "hunk."

It has been said that aphrodisiacs may well be the one cultural constant throughout human history; they cross all barriers of race, ethnicity, age, social status, and economics. Physicians and the Food and Drug Administration may dismiss the connection between the consumption of certain foods and an improved love life, but legend often trumps science. Folks believe what they want to believe, irrespective of whether or not it is in line with scientific reality. As a result, select foods and increased sexual activity seem to be a persistent cultural constant—a coexistence akin to other cultural pairs (horse and carriage, peanut butter and jelly, Chicago Cubs and never in the World Series).

> **FAST FACT:** In ancient times, it was traditional to present newly-weds with honey to help them enjoy their first sexual encounters, as it was said to intensify orgasms. Interestingly, though, there is no etymological connection between "honey" and the term "honeymoon." Several dictionaries report the etymology of "honeymoon" (originating in the sixteenth century) as "the idea that the first month (as determined by lunar phases) of marriage is the sweetest."

Part of the legend uniting certain foods and human sexual activity had its genesis in the apparent resemblance of some foods to human genitalia. From very ancient times the following items have been noted as potential aphrodisiacs: eggs, onions, celery, asparagus, bananas, and caviar (note the not so subtle references to human anatomy). Several ancient storytellers have described shucked clams as resembling miniature testicles, while oysters have been described by more than one philosopher as labial.

Additional stories from various parts of the world point to the reproductive capacities of selected plants and animals as evidence of their potential influence on human sexual activities. Ginseng, because of its striking resemblance to the human body, has also been tagged as an aphrodisiac. In fact, the word "ginseng" translates to "man root." Ripe pomegranates, for example, are filled with lots and lots of seeds. Eat pomegranates (so they say) and you, too, will be filled with a plethora of reproductive possibilities.

Those readers who live in western states will be familiar with a dish often appearing on many restaurant menus: "Rocky Mountain Oysters," or bull calf testicles coated in flour, pepper, and salt and deep fried. Eating calf testicles is supposed to enhance one's sexual performance because they contain testosterone. Unfortunately, however, cooking breaks down the testosterone. Nevertheless, the legend continues.

> **FAST FACT:** The World's Largest Rocky Mountain Oyster Feed is held in Eagle, Idaho, the second weekend in July.*

As we will learn in chapter 7, clams are considered by some to be reproductive superheroes among the world's animals. During select times of the year, female clams release millions of egg cells into the water. At the same time, males release dense clouds of sperm cells. Thus, as the logic goes, if I eat clams then I, too, will be able to produce dense clouds of sperm cells (or, if I were a female, millions of egg cells). What this legendary connection fails to mention, however, is that this reproductive act (among clams) takes up so much energy that it is often the last thing they do before dying.†

> **FAST FACT:** At various times throughout human history the following items have all been considered aphrodisiacs: rhinoceros horn, deer antler, cloves, sandalwood, ambergris (a.k.a. whale vomit), alder bark, okra, deer penis (self-explanatory), watermelon (yes, watermelon), horny goat weed (must be the name), gypsyweed rose petals, truffles, and, of course, chocolate. (Chocolate contains chemicals thought to affect neurotransmitters in the brain, which may heighten pleasure.‡)

Interestingly, the legendary connection between clams and sex got a scientific boost in 2005 when a group of Italian and American scientists released findings that amino acids found in bivalves (clams, oysters, mussels, scallops) have the potential to raise sexual hormone levels. It should be noted that the study was conducted on a Mediterranean species of

---

* Reporter Rachel Krause began a newspaper article (*Boise Weekly*, June 9, 2010) about this annual event with the following sentence: "Nothing says Eagle like a piping-hot plate of deep fried gonads." OK, I'm sold!

† Leading one to suspect that this act may be the origin of the phrase heard in singles bars every Saturday night, "I'd die for some sex right about now!"

‡ I have the great fortune to live less than thirty miles away from the chocolate capital of the world—Hershey, Pennsylvania. Now you know why I smile a lot!

mussels and demonstrated that those specific aminos—D-aspartic acid and N-methyl-D-aspartic acid—induced sexual hormone production in rats, not humans.

"The supposition for centuries was that oysters, clams, and mussels have been thought to have aphrodisiac properties," said researcher George Fisher, a professor of chemistry from Barry University, in Miami Shores, Florida. "And they were eaten raw for that purpose." But what he and his colleagues have discovered is that mussels, clams, and oysters contain compounds that have been shown to be effective in releasing sexual hormones, such as testosterone and estrogen. "We found there might be a scientific basis for the aphrodisiac properties of these mollusks," Fisher concluded.

On the other hand, Robert Shmerling, an associate professor of medicine from Harvard Medical School, countered, "The findings are certainly interesting, but we still have a way to go before saying that there is scientific evidence that clams, oysters, and scallops boost libido." As of the writing of this book, no follow-up studies have been conducted to determine the impact on humans, but millions of seafood lovers the world over are eagerly awaiting the possibilities!

I think we could safely say that the connection between clams and aphrodisiacs is a cultural constant. In short, eat more clams equals have more sex. Since clams are not a "regular" in the culinary field of vision for most of us, the lines around them often get blurred. Scientific reality often melts away, replaced by supposition, guesses, and dreams. In many ways, clams as aphrodisiacs may be more wish fulfillment than scientific reality. Dogs and cats we see every day; we can easily describe them with facts. Clams, on the other hand, are occasional visitors to our world, so their lives (and our desires) may be shaped more by conjecture than by reality. What often arises is a legend, a story, or a fictional narrative based on what might be, rather than on what is. But, isn't that part of the allure of clams in the first place?

Perhaps the famous Shawnee chief Tecumseh (1768–1813) said it best: "When the legends die, the dreams end."

Chapter 4

Clams as Culture

. . . the chowder being surpassingly excellent, we despatched it with great expedition.

—Herman Melville, *Moby Dick*

MOST OF MY SUMMERS AND MUCH OF MY (MIS-spent) youth took place along the beaches of southern California. Surfing, surf music ("Surfin' USA," "California Girls," "Surfin' Safari," "Surfer Girl"), and surfer duds (flip-flops, cut-offs) dominated much of the teenage culture of the sixties. It certainly dominated our language. Terms such as "woody" (an old station wagon with wood strips on the doors), "bitchin'" (cool or awesome), "hang ten" (to ride a surfboard with ten toes over the nose of the board), "stoked" (full of enthusiasm), "kick-out" (to make a controlled exit from a wave by riding up the face and over the top), and "wahine" (a girl surfer; or, should I say, a bitchin' surfer girl) frequently dominated our conversations. However, now that I'm a resident of landlocked Pennsylvania, I must admit that vocabulary seldom finds its way into my verbal exchanges and, hopefully, seldom pops up in my undergraduate lectures. Perhaps I've just learned to move on . . . culturally speaking.

> **FAST FACT:** Pennsylvania is the only one of the original thirteen colonies that does not border the Atlantic Ocean.

Cultural historian L. Robert Kohls defines "culture" as "an integrated system of behavior patterns characteristic of the members of a particular society."* That is to say, one group of people may behave or believe in ways different from another group. Those aspects of human expression may include language, technology, cuisine, art, music, social conventions, gender roles, holidays, social structure, and religion. In essence, one culture may embrace or define those aspects quite differently from another culture.

One of the unique characteristics of a culture is its cuisine. For example, my daughter and her husband live in northern England and the cuisine there is considerably different than it is here. Thus, when my wife and I visit England, we are treated to meals of bangers and mash, steak and kidney pie, fish and chips, toad-in-the-hole,† and Yorkshire pudding—not meals you would expect to find at one of the ubiquitous fast-food restaurants along the highways and byways of the United States. Just as in every other country, part of the English culture is the food typically consumed by its residents. So it is here, where our culture is partially defined by the burgers, fries, carbonated beverages, and the golden-arched places that sell them.

But culture is also an accumulation of experiences—a compendium of beliefs, ideas, practices, customs, habits, fashions, conventions, rites, and ceremonies shared by one group of people but not by another, or in the same way. The books that are written, the art that is produced, the food that is prepared, the music that is composed, and the icons that are celebrated may vary (and often do) from culture to culture. That is what makes Americans, the English, and every other culture in the world unique from each other.

FAST FACT: The Yanomamo tribe in South America has a particularly distinctive cultural tradition. The Yanomamo do not bury deceased individuals. Instead, when a person dies, his or her remains are cremated, and the ashes are subsequently mixed with bananas. The ash-and-banana mixture is then given to family members to be eaten. The Yanomamo believe that this practice helps preserve the soul of the deceased individual.

---

* L. Robert Kohls. *Survival Kit for Overseas Living* (Yarmouth, ME: Intercultural Press, Inc., 1996), page 23.

† In the eighteenth century this dish (sausages in Yorkshire pudding batter) was known as "pigeons in the hole." It seems that small, defenseless (and very cute) birds were used in place of sausages. Although the Brits are noted for eating some strange concoctions now and then, I suspect that this recipe never caught on to the same degree as, say, good old fish and chips.

My own research into the topic of clams revealed some interesting data relative to how clams have infiltrated our culture. While not predominant cultural figures, clams have influenced significant aspects of our lives in ways that many might find extraordinary. For example, clams have figured prominently in literature, musical compositions, and artistic ventures. And, of course, they have long held influence over various culinary adventures. So, let's examine some of the cultural representations of clams—specifically in the United States.

Literature

"Call me Ishmael."

Those three words may be the most recognizable opening line in American literature. Say them out loud in any crowded room and at least 99 percent of those present will recognize the words that open the novel *Moby Dick* by Herman Melville. First published in London in October 1851 and then in New York in November 1851, *Moby Dick* (despite its initial lukewarm reviews) has gone on to become one of the classics of American literature—a novel that is not only required reading in high school English courses across the country, but one that has been turned into at least a half-dozen motion pictures. It has also generated hundreds of critical reviews, doctoral dissertations, and articles in all manner of various periodicals. It could be safely said that *Moby Dick* is a staple of literature, the benchmark against which all other novels generated in the first century of this country's existence have been measured. According to at least one expert, *Moby Dick* is at the center of the canon of American novels.*

As you may recall, the story recounts the adventures of the sailor Ishmael, specifically his voyage on a whaling ship. *The Pequod*, commanded by the obsessively demented Captain Ahab, sets sail in search of a mysterious beast, the enigmatic and elusive Moby Dick, an albino sperm whale lurking in the abyss of the Pacific Ocean. Ahab is driven by a previous encounter with the whale, one in which his ship was destroyed and his leg severed. As the story progresses, Ahab's anger and obsession with the great white whale consumes him. His singular and ultimate goal is to search out his submarine enemy and destroy it any way he can, even if it means taking his entire crew with him in the process.

---

* Harold Bloom. *The Western Canon: The Books and School of the Ages.* (New York: Harcourt Brace & Company, 1994).

While the story is focused on the search for the elusive whale, there is one chapter that celebrates clams. Chapter 15 (appropriately entitled "Chowder") tells of an evening when Ishmael and his partner, Queequeg, in want of some good food, come upon Try Pots—an eatery of some renown, noted for its clam chowder. Herewith is an excerpt from that chapter:

It was quite late in the evening when the little Moss came snugly to anchor, and Queequeg and I went ashore; so we could attend to no business that day, at least none but a supper and a bed. The land-lord of the Spouter-Inn had recommended us to his cousin Hosea Hussey of the Try Pots, whom he asserted to be the proprietor of one of the best kept hotels in all Nantucket, and moreover he had assured us that Cousin Hosea, as he called him, was famous for his chowders. In short, he plainly hinted that we could not possibly do better than try pot-luck at the Try Pots.

. . . . .

And so it turned out; Mr. Hosea Hussey being from home, but leaving Mrs. Hussey entirely competent to attend to all his affairs. Upon making known our desires for a supper and a bed, Mrs. Hussey, postponing further scolding for the present, ushered us into a little room, and seating us at a table spread with the relics of a recently concluded repast, turned round to us and said—"Clam or Cod?"

"What's that about Cods, ma'am?" said I, with much politeness.

"Clam or Cod?" she repeated.

"A clam for supper? a cold clam; is that what you mean, Mrs. Hussey?" says I, "but that's a rather cold and clammy reception in the winter time, ain't it, Mrs. Hussey?"

But being in a great hurry to resume scolding the man in the purple shirt who was waiting for it in the entry, and seeming to hear nothing but the word "clam," Mrs. Hussey hurried towards an open door leading to the kitchen, and bawling out "clam for two," disappeared.

"Queequeg," said I, "do you think that we can make a supper for us both on one clam?"

However, a warm savory steam from the kitchen served to belie the apparently cheerless prospect before us. But when that smok-ing chowder came in, the mystery was delightfully explained. Oh!

sweet friends, hearken to me. It was made of small juicy clams, scarcely bigger than hazel nuts, mixed with pounded ship biscuits, and salted pork cut up into little flakes! The whole enriched with butter, and plentifully seasoned with pepper and salt. Our appetites being sharpened by the frosty voyage, and in particular, Queequeg seeing his favourite fishing food before him, and the chowder being surpassingly excellent, we despatched it with great expedition: when leaning back a moment and bethinking me of Mrs. Hussey's clam and cod announcement, I thought I would try a little experiment. Stepping to the kitchen door, I uttered the word "cod" with great emphasis, and resumed my seat. In a few moments the savoury steam came forth again, but with a different flavor, and in good time a fine cod-chowder was placed before us.

We resumed business; and while plying our spoons in the bowl, thinks I to myself, I wonder now if this here has any effect on the head? What's that stultifying saying about chowder-headed people? "But look, Queequeg, ain't that a live eel in your bowl? Where's your harpoon?"

Fishiest of all fishy places was the Try Pots, which well deserved its name; for the pots there were always boiling chowders. Chowder for breakfast, and chowder for dinner, and chowder for supper, till you began to look for fish-bones coming through your clothes. The area before the house was paved with clam-shells. Mrs. Hussey wore a polished necklace of codfish vertebra; and Hosea Hussey had his account books bound in superior old shark-skin. There was a fishy flavor to the milk, too, which I could not at all account for, till one morning happening to take a stroll along the beach among some fishermen's boats, I saw Hosea's brindled cow feeding on fish remnants, and marching along the sand with each foot in a cod's decapitated head, looking very slipshod, I assure ye.*

FAST FACT: Although this novel is named for, and focuses on, the Great White Whale, Moby Dick only appears in three of the 135 chapters of the book.

---

* Herman Melville. *Moby Dick.* (London: Richard Bently, 1851).

In 1906 author William John Hopkins penned a novel about the life and times of a professional clammer. *The Clammer*, while not attaining the stature and accolades of *Moby Dick*, was a work of fiction that entertained turn-of-the-century readers with the story of Adam, a poor clammer who falls in love with the rich man's daughter, a charming young lady known as (yup, you guessed it) Eve. Not only does he win over the young lady, he also shares his passion for clams with his future father-in-law. The rich man quickly becomes a "clam convert" as indicated in the following passage:

> And Old Goodwin, after further searching in the tree, drew forth a clam hoe and a basket; and being thus equipped, he hied them to the flats, which were, by now, almost bare, and he began to dig. Now that is a luxury which the rich may seldom have, that they should dig for clams. Old Goodwin enjoyed it mightily, splashing here and there in his boots, and digging as the fancy seized him; which was as like to be where the clams were not as where they were. But he cared not at all, and drew long breaths for very joy of living; and the clams that he found he put within his basket.

Not only does the author satisfy his audience's literary appetites; he presents readers with aspects of the clammer's life that stimulate their culinary appetites, too. Witness the following description:

> The hole was scooped in the ground and lined with great stones. And on those stones I kindled a fire that roared high; and when it had burned long and the stones were hot, I raked the ashes off. Then I shook down upon the stones fresh seaweed from the pile, and on the seaweed laid the clams that I had digged, myself—and alone—that morning. Then, more seaweed; and the other things, in layers, orderly, with the clean, salt-smelling weed between: the lobsters, green and crawling, and the fish, fresh caught, and the chicken, not too fresh, and the sweet and tender corn, and sweet potatoes. And over all I piled the weed and made a dome that smoked and steamed and filled the air with incense.*

The Clammer was sufficiently successful that the author penned a follow-up novel entitled *The Clammer and the Submarine* (1917). The "submarine" was not, as some might expect, a veiled reference to Moby Dick,

---

* W. J. Hopkins. *The Clammer*. (Boston: Houghton Mifflin, 1906).

but rather a convoy of World War I war ships cruising near Adam's beloved clam beds.

Melville and Hopkins certainly provide evidence that clams have had some degree of literary value and presence in American literature.

Language

From the time of the ancient Greeks (and perhaps even earlier), the connection between language and culture has been firmly cemented. That is to say, the shared language of a community may be one of the most significant "markers" that identifies and clarifies that community. In other words, understanding the language of a community helps us understand the culture of that community.

Evolutionary anthropologist Robin Dunbar has proposed the concept that language evolved as early humans began to live in large communities— communities that required the use of complex communication to maintain social coherence and permanence. By the same token, each community identifies itself by its language and, at the same time, distinguishes itself from every other community by the various ways in which it uses its language.

Take, for example, the way English is spoken in countries as diverse as the United States, England, Ireland, Canada, Australia, and New Zealand. If you have traveled to any English-speaking countries outside the United States, you know that people in those countries use English somewhat differently than we do in the States. That's part of their culture, their identity, and what distinguishes them from every other culture.

For example, when I visit my daughter in England, I have to constantly remember that the British speak a form of English considerably different than I am used to in south-central Pennsylvania, where I live. Yes, it's English, but it's a form of English replete with terms, colloquialisms, and euphemisms that are born of, and distinctive to, the English culture—quite different from the culture I grew up in southern California and certainly different from the one in which I now reside.

FAST FACT: The history of the English language is usually divided into three main periods: Old English (450–1100 AD), Middle English (1100–1500 AD) and Modern English (since 1500). Over that span of time, the English language (in England) has been influenced by several other languages including German, Celtic, Latin, Old Norse, Old French, and Greek.

It hardly goes without saying that we, in the United States, also have terms, words, phrases, colloquialisms, idioms, slang, and language that are distinctive and unique—vocabulary quite different from our Anglo-Saxon relatives. That is certainly true when we include clams in the discussion. Herewith are a few of the special "clam-isms" we use in everyday conversations and language.

"Happy as a clam": This means to be overwhelmingly happy or content. ("I'm as happy as a clam since I got that big raise.") The full and complete phrase is actually "happy as a clam at high tide." It refers to the fact that clams are usually dug up during periods of low tide. When the tide is high, then the clams are covered and safe from any near-by clammers. This, as you might imagine, would make them (if they were to have human emotions) very happy, indeed.

"Clam up": This refers to someone who has stopped talking, either voluntarily or involuntarily. In most cases it refers to the sharing of information. ("She suddenly clammed up when the detective asked her about her former boyfriend."). Since clams spend a good deal of their life with their two shells in a closed, or nearly closed, position, clamming up is similar to when humans keep their lips together or their mouths shut—they don't (or can't) say anything.

"Shut up like a clam": This has the same meaning as "clam up." ("He shut up like a clam when asked about the location of the stolen painting.")

"Clamshell" (container): "He couldn't finish his meal, so the waitress gave him a clamshell container so he could take his leftovers home." Clamshell containers are used by many restaurants and eating establishments for the convenience of customers. Made from Styrofoam®, they open and shut just like the two shells on a clam. They often have a self-locking tab to prevent any accidental opening. They are used primarily as a transport container until the customer returns home and can place the box into the refrigerator. Not meant for long-term storage, they are a convenient way to carry leftover food from one location (restaurant) to another (home). This assumes, of course, that we actually remember to take the clamshell container from the restaurant when we leave. The term can also be applied to containers for many other items, including certain laptop computer cases, cosmetics containers, and CD cases.

"Clammy hands": When people get excessively nervous, say just before a job interview, their hands tend to sweat a little more than usual. This is a normal physiological reaction, but also one subject to ridicule or condemnation. ("He would always get clammy hands just before he asked a woman out on a date.") Clammy is etymologically related to a root word of clam, which originally meant sticky. (The clamshells were stuck or pressed together.)

"Clams" (currency): The term can be a slang term for "dollars." ("That stupid car cost me 25,000 clams!") Native Americans living in the eastern woodlands of North America used wampum to commemorate special ceremonies and historical events, or as something to exchange during events such as marriages. Wampum was also used by European colonists as a form of currency in their various trades with Native Americans.

Wampum usually consisted of beads made from channeled whelk shell or hand carved from quahog clamshells. Although they weren't used as currency by Native Americans, Europeans frequently used wampum belts as currency throughout the seventeenth century. Sometime in the early eighteenth century their use as a form of currency lost favor, most likely because of an epidemic of red tide that virtually wiped out the entire whelk and quahog population. Nevertheless, even though the beads made from clams became significantly less popular, the slang use of "clams" for money has remained a part of our culture ever since.

"Clam" (musical): Sometimes it is a slang term for a mistake during a musical performance. ("The musician kept recording his song in spite of the two clams by the keyboardist.") Many churches in Europe signaled services by ringing a series of bells in a bell tower. Usually, one person would be in charge of ringing several different bells. Occasionally, a mistake would be made, and two bells would be rung simultaneously. This was known as a "clam" (like two shells closing simultaneously).

I also discovered at least one reference (though unsubstantiated by other documents) suggesting that the use of this term eventually resulted in our word "clamor," a term referring to a jumbled collection of simultaneous noises or various voices. As a result, it seems possible that a "clam" would be an appropriate term for a musician's boo-boo or flub in a musical performance.

Suffice it to say, the word "clam" has been (and will continue to be) a multi-functional word throughout the (American) English language. Used synonymously with money, noise, and feelings, it has evolved into any number of linguistic uses and will probably continue to do so for some time in the future. The same cannot be said of their bivalve cousins; for example, imagine the following story line: "Big John tightened his meaty hands around Benny's neck and squeezed even harder. 'Listen, punk,' he hissed in Benny's ear, "You get me the thirty thousand *oysters* by noon or you're a dead man!" Somehow, it just doesn't cut it!

> **FAST FACT:** Pelecypodophobia is the fear of bivalve mollusks. Just thought you'd want to know!

Art

Much of how we define culture is determined by art. That is to say, how art is expressed and how art is supported varies from culture to culture. We praise the art of Renaissance painters such as Leonardo da Vinci, Raphael, and Michelangelo. Often, when we think of ancient Greek culture, we think of some of its dynamic and classical art. Greek statues, paintings, and architecture are celebrated for their detail and expression. And we marvel at the intricacies of long-ago artists who decorated their caves with antelopes and bison, who created exquisite sculptures in long-forgotten Central American valleys, and who carved towering totems along the vibrant streams and meandering rivers of the Pacific Northwest. Their art helps define the cultures to which they belonged—it helps define the people.

Clams have been a subject of artists for hundreds, if not thousands, of years. Clams can symbolize reproductivity, occupation, hobby, simple beauty, femininity, and cuisine. In fact, during the Italian Renaissance the clamshell symbolically represented the vulva and matrix (womb) as life-receiving and life-giving organs. It was widely used in baroque emblems to symbolize Mary and the divine conception of Jesus.

That clams (and their shells) have been embraced by so many different artists from so many different times is a testament to their enduring artistic elements . . . and their role in so many aspects of our lives. Let's take a look at a few representative paintings from different eras.

Note the clamshell design behind Mary's head in the painting *Madonna and Child* (1440–1445) by Fra' Filippo Lippi.

The Birth of Venus by Sandro Botticelli (circa 1486)

The Birth of Venus by Botticelli.

Suffice it to say, this painting may be one of the most recognizable in all of Renaissance art. It's been used in advertisements and has been replicated and duplicated in numerous prints ever since its original creation in 1486. It was commissioned by the Medici family, and the original currently hangs in the Uffizi Gallery in Florence, Italy.

The painting depicts the naked goddess Venus emerging from the sea as a full-grown woman. It is believed that the model for Venus was Simonetta Cattaneo de Vespucci, a most beautiful woman and, according to current thought, a great love of Botticelli. (He expressed a desire to be buried at her feet when he died.)

The painting depicts the arrival of Venus on an enlarged clamshell, one which has been blown toward shore by Zephyrus, god of the west wind, and by the gentle breeze Aura. Look closely and you will notice a most striking resemblance between Venus and Aura—an indication, perhaps, of Botticelli's fascination with his model. To the left of Venus is one of the Horae, goddesses of the seasons, who waits to receive the goddess of love and holds a flower-covered robe in anticipation of Venus's arrival.

The clamshell occupies the front center of the painting and, as such, much of the focus is on this item. Renaissance painters often used seashells in their work as a metaphor for a woman's vulva. In Botticelli's painting, he places the goddess of love on that symbolic seashell and allows the wind to gently blow this detailed symbol across the water. Thus, Venus is symbolically connected to her eventual human "assignment": the promotion of love, beauty, enticement, seduction, sex, fertility, and prosperity.

At the time, this painting was considered to be extremely pagan, since most of the artworks composed by Boticelli's contemporaries portrayed religious themes as opposed to classical themes. (Interestingly, many of Boticelli's other "pagan" works were burned and destroyed. This painting, however, escaped that destruction.)

FAST FACT: Adobe Illustrator used a stylized representation of "The Birth of Venus" in its splash screen through Version 10.

A Basket of Clams by **Winslow Homer (1873)**

Celebrated American artist Winslow Homer (1836–1910) was enamored of the sea. In June of 1873 he traveled to Gloucester, Massachusetts, where he first began experimenting with watercolors. It was during his time in Gloucester that he focused on the play, actions, and activities of children—depicting them sitting on wharves, playing in the boats, doing their chores, or simply engaging in various summertime activities.

One of the paintings that emerged from that summer was "A Basket of Clams"—an engaging portrayal of two boys sauntering along a shell-strewn beach carrying a basket of clams between them. The details of this painting are particularly noteworthy: the two-masted sailboat just behind the boys, typical Gloucester buildings in the background, and seashore detritus haphazardly scattered on the beach. The play of light across these various elements is typical of Homer's paintings and his celebration of the postbellum era in American history. This painting is currently on display at the Metropolitan Museum of Art in New York.

A Basket of Clams.

Clam Shell by Georgia O'Keeffe (1930)

Although American artist Georgia O'Keeffe is best known for her work at the Ghost Ranch in New Mexico, she also spent considerable time in Maine focusing on topics related to seashore life and the ocean. A master of shapes and colors, she produced a series of paintings depicting clam shells in various stages of opening and closing. This series included *Open Clam Shell, Closed Clam Shell,* and *Slightly Open Clam Shell.*

As with most of her clamshell paintings, there is a sense of simplicity and design. If you were to examine them carefully, you'd note that the colors are in shades of silver and alabaster. Further examination would also reveal a vertical line broken at midpoint by a circle. These elements were features in many O'Keeffe paintings, including work depicting natural history subjects, such as flowers and corn.

Viewing these paintings, it is clear that the imagery is both overt and distinct. Each of the clamshell paintings can be viewed as a self-portrait, a depiction of a woman (once again, the clam is symbolically vulval) in the process of shutting down. Look inside the clamshell and you will see a void, an empty chamber. Later in her life, O'Keeffe was heard to comment, "I find that I have painted my life—things happening in my life—without knowing."

> **FAST FACT:** Georgia O'Keeffe has been recognized as one of the most compelling American artists of the twentieth century. In 1977 she received the Presidential Medal of Freedom from President Gerald Ford. In 1985 she was presented with the National Medal of the Arts by President Ronald Reagan.

Treading Clams, Wickford by William James Glackens (1909)

William Glackens (1870–1938) was known as an American realist painter. Prior to the start of World War I, he focused on everyday street scenes and depictions of daily life in New York and Paris. His work featured dark

Treading Clams, Wickford.

colors and vibrant details. Later, after the war, his work was strongly influenced by Renoir and became brighter in both tone and composition.

Although raised in Philadelphia, he eventually traveled to Europe with several other artists to experience the "Continent" and to be influenced by European techniques. He was particularly enamored of the Impressionists and Post-Impressionists, and much of his later work clearly reflects that influence. Upon his return to America in 1896, he began to work as an artist for several newspapers and periodicals. By 1910 he was fully engaged in Impressionistic art and he was often compared to one of his artistic heroes, Renoir. (He was frequently referred to as "The American Renoir.") In his later years, he focused primarily on landscapes, most notably beach scenes. *Treading Clams, Wickford* is a prime example of that period of his life.

Quahog on Clams by Dominic White (2011)

Dominic White is a contemporary artist living in Maine. His paintings reflect his New England background and "Down East" influence. Titles of his work include *A Fortier Docked in Maine, Blueberries, White Wine and Lobster,* and *The Light Keeper's House.* He has explored various mediums and styles in his work with a current focus on drawing.

FAST FACT: In Pismo Beach, California ("Clam Capital of the World"), there are several six-foot-tall cement clams placed around town. Before each holiday, the clams are painted and dressed up to resemble familiar figures: Easter Clam, Turkey Clam, Santa Clam, Leprechaun Clam, etc.

Songs

As you may recall from the introduction to this book, clams do not suffer from lack of musical exposure. If singers such as Elvis saw fit to extol the virtues of a dance inspired by the imaginary actions (or inactions) of clams, then it would seem logical that other musicians (hoping to cash in on mollusk-inspired melodies) should also step into these royalty-rich waters with their own clam concertos or (in some cases) clam cacophony. Here are just a few examples of how clams have entered our culture via a musical portal.

"Do the Clam" by Elvis Presley

If, as a teenager, you were grabbing someone near to you, and then engaging in a sufficient amount of turning and teasing and hugging and squeezing, you may well have been doing "The Clam." Do all those actions on a beach and, as Elvis would say, "That's all right!" Since "Do the Clam" was sung by "The King of Rock and Roll," this song may have become the best-known clam tune ever recorded . . . if not remembered.

> **FAST FACT:** When it was released as a single in 1965, "Do the Clam" reached #21 on Billboard. Apparently, Australians were "doing the Clam" more than Americans, as the song rocketed to #4 on their charts.

"Clam Chowder" by The Sadies

This short instrumental (one minute and thirty-five seconds) from their debut album (*Precious Moments*) in 1998 identified the Toronto-based Sadies as a distinctive new Canadian group. Their sound—a compilation of traditional country, surf music, and garage rock—signaled their emergence as a singular rock group in a very competitive field.

"Clams Have Feelings Too (Actually They Don't)" by NOFX

At the very least, the lyrics to this song leave listeners scratching their collective heads. It's not that they're bad, it's just that they're . . . well . . . meaningless. For example, the singers tell listeners that clams should really be smiling (remember "happy as a clam"?), but they really can't, since they don't have a face. The group then launches into a high school biology lesson informing the audience that these bivalves lack basic anatomical features such as ears, eyes (which prevents them from shedding tears), and a spinal cord. As a result, all clams should be summarily consumed, since they essentially don't have any sensations or, quite obviously, emotions. I suspect that the attempt to juxtaposition invertebrate zoology and bad rock and roll forever consigned this song to the category of "Most Forgettable Music Ever Made."*

---

* Sending the parents of those teenagers into all sorts of paroxysms.

"Acres of Clams" by Pete Seeger

This song was originally penned by Francis D. Henry in 1874 (or thereabouts). The song has also appeared under several other titles including "Lay of the Old Settler," "Old Settler's Song," and its more well-known appellation, "Acres of Clams."

The song is frequently sung to the tune of "Old Rosin the Beau," and at one time (so the rumor goes) it was believed to be the state song of Washington. Radio personality Ivar Haglund, a DJ in Seattle, Washington, used the tune as the theme song for his popular radio program. (He also used the song's title for his signature restaurant, Ivar's Acres of Clams.) Later, it was revealed that Pete Seeger and Woody Guthrie taught the song to Haglund, who liked it so much that he adopted it as his own. Pete Seeger eventually recorded the song for his album *Sing-a-Long at Sanders Theatre, 1980*. This is an excerpt from the original folk song:

> *I've wandered all over this country,*
> *Prospecting and digging for gold,*
> *I've tunneled, hydraulicked and cradled,*
> *And I nearly froze in the cold.*
> *And I nearly froze in the cold,*
> *And I nearly froze in the cold,*
> *I've tunneled, hydraulicked and cradled,*
> *And I nearly froze in the cold.*
>
> . . .
>
> *No longer a slave of ambition,*
> *I laugh at the world and its shams,*
> *And I think of my happy condition,*
> *Surrounded by Acres of Clams,*
> *Surrounded by Acres of Clams,*
> *Surrounded by Acres of Clams.*
> *And I think of my happy condition,*
> *Surrounded by Acres of Clams.*

"Surf Clam!" by The Mel-Tones

In 2005, the musical group The Mel-Tones came out with an album entitled *Surf Sensation*. This album is a collection of thirty-three songs mined from the *Spongebob Squarepants* TV show. The last song on the album is the instrumental "Surf Clam."

> **FAST FACT:** In 2013, *TV Guide* ranked *Spongebob Squarepants* as the eighth "Greatest TV Cartoon of All Time." (*The Simpsons* was ranked #1.) Too bad none of the main characters are clams!

"Poison Clam" by The Phantom Surfers

The Phantom Surfers were a surf band originally formed in 1988. In 2000 they issued an album entitled *A Decade of Quality Control 1988–1999.* Track #15 on that album is the instrumental "Poison Clam." You can download the song (for less than a dollar) online; but, trust me, I don't think your life will be significantly affected should you decide not to listen to this tune. (Maybe it's just me, but I can't figure out the connection between discordant guitar strumming and venomous clams.)

Movies

Diggers (2006) is the only movie with clams as major characters (at least tangentially). The story is a coming-of-age tale about a group of four friends on Long Island. These thirty-somethings seemed consigned to a mundane life of clamming—just as their families have been for generations. They try to come to grips with what appears to be a bleak (and never-ending) existence as clammers while dreaming of a better life somewhere else.

One reviewer on *Rotten Tomatoes* (www.rottentomatoes.com) said this about the film, "*Diggers* bumps along from trauma to trauma like a clam-digging boat on choppy waters." Another commented, "*Diggers* is all shell and no neck and belly." After watching the film, I suspect both those reviewers were being a bit too generous, although I did appreciate their clam-related language.

The film starred Paul Rudd, Maura Tierney, Lauren Ambrose, and Ron Eldard. It was directed by Katherine Dieckman, and the screenplay was by Ken Marino.

Clams, in all their forms and appellations, seem to be firmly established in the American psyche, if not the American culture. We use them in our food, our language, our writings, our entertainment, our songs, and our art. They are part and parcel of our past and of our present. That they have entered so many facets of our lives is a testament to our endearing and lasting (not to mention culinary) admiration of these creatures as well as their singular ability to burrow their way into multiple aspects of our existence. As far as American culture is concerned, clams are heralded and celebrated. You and I should be so lucky!

PART II

Nomenclature and Biology

C h a p t e r 5

By Any Other Name

*What's in a name? That which we call a rose by any other name
would smell as sweet.*

—William Shakespeare, Romeo and Juliet

ALL THINGS CONSIDERED, PERHAPS THE ONE thing you hold nearest and dearest to your heart is your name. Your name is you! Your name goes on every job and college application, every piece of governmental paperwork, and all manner of Christmas cards, text messages, love letters, and refinancing loans. Your name is your identity—it separates you from every "Tom, Dick, and Harry" (as well as every "Emma, Abigail, and Olivia") in the world and becomes your most personal calling card from birth until well after your passing.*

In so many ways, your name is your personality, your essence, your identification in the world. It is as much who you are as anything you may own. Basically, your name characterizes you—culturally, socially, professionally, and academically—now and forever. Or, as the Romans used to say—in Latin no less—*nomen est omen*, or name is destiny.†

---

* A 2006 study published in the *Proceedings of the National Academy of Sciences* claims that humans may not be the only animals who use personal names. Researchers from the University of North Carolina Wilmington studied bottlenose dolphins in Sarasota Bay, Florida. They cite convincing evidence that the dolphins had names for each other. Apparently, a dolphin chooses its name as an infant.

† In 2012, researchers at Humboldt University in Berlin, Germany, discovered that dating website profiles with "unattractive names" are visited less often than those with "attractive

As I'm sure you're aware, your name (whatever it may be) most likely has a meaning, a symbolism, or a definition. Many people have names with roots in other countries, other cultures, and other languages. As a result, the definitions of those names may be quite common or quite unusual. Some meanings are biblical in nature, while others may be political or cultural. Still others may have names with no intrinsic meaning—the name was crafted by parents as something quite different from every other name in the world, or it was the pronunciation, more than any particular meaning that was more important. It is a given that many new parents-to-be will spend enormous lengths of time (and a great deal of Internet searching) before they select a baby's name. They want to get both the sound and the meaning absolutely perfect.*

However, if you're planning a new addition to the family you might want to reconsider some of the names below. They certainly sound interesting, but their meanings are something else.

Name	Meaning
Bernard	"brave bear"
Bethany	"house of figs"
Brody	"muddy place"
Byron	"place of the cow sheds"
Cade	"round, lumpy; barrel-maker"
Caleb	"dog"
Cameron	"crooked nose"
Chelsea	"landing place for chalk"
Isaac	"he will laugh"

---

names." "And, which names are most unattractive?" you might be asking (as did I). For males, the names Kevin, Justin, Marvin, and Dennis were less favorable; for females, Mandy, Chantal, Celina, and Jacqueline were least favorable. The most favorable names for males were Jacob and Alexander; for females, Charlotte and Emma. Thus, it seems that if you aren't getting "enough action" on your own dating site, you might want to consider changing your name (or, at the very least, using a *nom de plume*). Wouldn't you agree, Marvin?

* My name, Anthony, is of unknown Etruscan origin. (In short, nobody knows what it means.) However, in the third century there was Saint Anthony the Great; probably not a relative, but, then again, who's to say?

Jennifer*	"fair phantom"
Julian	"down-bearded youth"
Landon	"long hill"
Logan	"hollow"
Mallory	"luckless"
Maria	"bitter"
Paige	"servant"
Travis	"toll collector"

When you go to work tomorrow, take a close look at the people you work with. You may discover that your best friend is a "place of cow sheds," you may sit next to a "crooked nose" in the company cafeteria, or you may share a drink with a "down-bearded youth" at the annual Christmas party. If so, I would respectfully suggest you keep your nomenclature knowledge to yourself. I guess some things are just better left unsaid—especially around the office!†

Now, let's consider clams . . . or more specifically, the names of clams. You might think naming clams would be relatively simple. A clam is a clam is a clam. Apparently, not so! Throughout history, clams have been saddled with any number of names that are not only confusing but downright frustrating (particularly for certain authors trying to write a book about clams). More often than not, the differences are regional. For example, you may order littleneck clams at a restaurant in Portland, Maine, and I may order littleneck clams in another restaurant in Portland, Oregon, and we discover that we each receive two completely different kinds of clams.

> **FAST FACT:** Portland, Oregon, was named after Portland, Maine. Its original name, however, was Stumptown.

---

* Several years ago, one of my undergraduate courses had twenty-one students. Five of the fourteen females in that section were all named Jennifer. Early in the course, I briefly considered calling every female in the class "Jennifer," figuring that my chances were pretty good (36 percent) of getting a female student's name correct. I didn't do that, but by the end of the semester, I still hadn't figured out which Jennifer was which.

† Be glad you don't work in an office full of folks from the Machiguenga tribe of the Amazon. Members of that tribe have *no* personal names. Sending interoffice e-mails would pose a truly unique problem, don't you think? And, let's not even talk about Facebook!

This clam-naming situation is further compounded by the fact that there are thousands of clam species throughout the world, of which more than 150 species are edible. It's those edible clams that are causing all the identification problems (at least for me). That's often because a particular clam's name may be determined by two significant factors:

- Where you are
- What form of a clam you're eating (or digging)

However, since clams have no emotions and probably wouldn't care one way or another what they were called, it behooves us as humans to try to get this whole nomenclature thing right. This is not an easy task simply because (as humans) we have certain customs, traditions, and "rules" we follow to make sure that we can communicate with the folks with whom we live. In clam talk, you may be referring to "gapers" and I may be referring to "horsenecks" (even though "gapers" and "horsenecks" are the exact same animal—*Tresus capax*).

In other words, you say "To-MAH-to" and I say "To-MAY-to."

So, let's see if we can put this whole clam name thing in some kind of logical arrangement once and for all. By the way, all the descriptions below apply to edible United States clams. Those readers from England or Australia or Canada* will have to locate another book to investigate the varieties of clams in your respective countries. Sorry.

The initial part of our nomenclature conundrum is easy (it does get slightly more confusing after this). First, all the commercial clams available in the United States fall into one of two very broad categories: hard-shelled clams (*Mercenaria mercenaria*) and soft-shelled clams (*Mya arenaria*). Let's take a look at each of those two species.

Hard-Shell Clams (*Mercenaria* genus)

Hard-shell clams can be found on both coasts. They are quite common throughout the eastern seaboard from the Gulf of St. Lawrence, down around Florida, and along the Gulf Coast to Texas. Several varieties inhabit the west coast as well. Hard-shells are mainly found in bays and estuaries

---

* Aussies in the mood for some edible bivalves (in stores or in restaurants) will order vongoles, pipis, dosinias, cockles, or razor fish. In England, sand gapers are the preferred bivalve. Canadians call their clams . . . well, "clams." Fortunately, our northern neighbors like to keep their seashore nomenclature nice and simple. I, for one, appreciate that.

from the intertidal zone to depths of about sixty feet. Although they can tolerate a wide variety of substrates, they tend to prefer sandy bottoms.

FAST FACT: Native Americans prized the shells of hard-shell clams for making wampum beads—a form of currency used by Europeans in the early years of this country. As a result, these clams were given the Latin name *Mercenaria*, which means "commerce," "wages," or "reward."

It's often been said that you can't judge a book by its cover. However, that's not the case with hard-shell clams. These clams have shells that are heavy, thick, and strong. The exterior of the shell is frequently grayish, often with a tinge of brown or tan. Look carefully and you'll also notice numerous concentric growth rings following the shape of the shell. The interior, as with many shells, is pure white, sometimes with a splotch of purple.

But, now, things start to get a little confusing. You see, hard-shell clams are grouped according to size. While there is no absolute agreement on these designations (different regions of the country have slightly different interpretations), the following "clam grades" are typical:

Group Name	**Size**
Buttons	< 1⅞ inch
Littlenecks	1⅞–2⅛ inch
Topnecks	2⅛–2⅜ inch
Cherrystones	2⅜–3⅛ inch
Chowders	> 3⅛ inch

Interestingly, if left alone, adult hard-shells can frequently grow as large as five inches across. They will also live for about twenty years, although specimens as old as fifty years have been reported.

Are you still with me? Because, now it's time to regionalize our hard-shell clams. That is to say, give them names based on where they live.* Since there are several different types of hard-shell clams found along the

---

* Just like we do. For example, folks who live in Nebraska are called Cornhuskers. Individuals who come from Wisconsin are known as Cheeseheads. And people who grew up in Southern California are known as "Surfer Dudes."

coastlines of the United States, I'm going to separate those clams into two distinct groups—clams commonly found along the east coast, and clams that typically take up residence along the west coast.

East Coast Hard-Shells

Quahog (*Mercenaria mercenaria*)

When you're talking about hard-shell clams on the east Coast, you're talking about quahogs, a word borrowed from Native Americans. Back in the old days, the Narragansett Indian word for clams was "poquahock." Early American settlers had a difficult time pronouncing that word and so, changed it to "quahog," a term still in use today. Interestingly, "quahog" has a number of pronunciations, depending on your accent: KO-hog (the preferred), KWO-hog, and KWA-hog. In much of the United States, quahogs are simply called "hard clams" or "hard-shell clams."

Just to keep things interesting (or more confusing), there are actually two separate species of quahogs: the northern quahog (*Mercenaria mercenaria*)* and the southern quahog (*Mercenaria campechiensis*). *Campechiensis* refers to the Campeche region of Mexico, where this specimen was first found and described in the late 1800s.

The northern quahog is distinguished by a heavy, dirty-gray to whitish oval shell that is relatively thin and smooth. The interior often has a light-purple shading along the outer margin. The southern quahog has a shell with a similar exterior color. However, the shell's interior is white and near the hinge are three well-developed grooves, called "teeth."

Like many species of trees, a quahog's age can be determined by counting its growth rings. Those rings are prominently displayed on the outside of its shell. Interestingly, as quahogs get older, they tend to grow more slowly (as do the senior citizens of many other species), so the growth rings get very close together. As a result, it is sometimes difficult to count them accurately. Yet, by counting those rings researchers have determined that the largest quahogs (four inches or more in length) are often as much as forty years old.

Although quahogs can be found along the North American Atlantic coast from Canada's Gulf of Saint Lawrence to Florida, they are particularly abundant between Cape Cod and New Jersey. Farther north, most waters are too cold for quahogs, restricting them to just a few relatively warm coves.

---

* The folks naming this species apparently thought this clam's Latin derivation was so cool that they repeated it. It's sort of like calling someone the same thing two times, like "stupid idiot, stupid idiot" for the guy who insists on tailgating you at seventy-five miles per hour on the freeway.

> **FAST FACT:** Our smallest state, Rhode Island, is situated right in the middle of "quahog country." As a result, it typically supplies a quarter of the nation's total annual commercial quahog catch.

West Coast Hard-Shells

Geoduck (*Panopea generosa*)

According to ancient folklore, extra large clams, known as geoducks (pronounced "gooey ducks"), are considered to be a natural aphrodisiac. If this is the first time you've seen a geoduck (below), then you may have sucked in your breath, bulged your eyes, and said to yourself, "My, oh, my, they look just like an . . . an . . . an aphrodisiac."

A geoduck from the state of Washington.

FAST FACT: The geoduck is the official mascot of Evergreen State College in Olympia, Washington. In a presumably symbolic reference to the geoduck's phallic appearance, the school's Latin motto is *Omnia Extares!* ("Let it all hang out!")

Geoducks are considered culinary treats, nay a culinary nirvana, in China. For many, they are the epicurean zenith of seafood. The large, meaty siphon is often prized for its briny sweetness and crunchy texture. Frequently served at Chinese celebrations and banquets, geoducks may be baked, fried, boiled, sautéed, or served as sushi, but are most popular blanched in a boiling broth. When you consider that these delicacies may sell for $150 a pound or more in a Chinese restaurant (You must admit, that's a lot of "clams" for a clam), and that the demand often outstrips the supply, you can understand why chefs (and the owners of restaurants where those chefs work) love this marine creature.

The name for these clams is most unusual. It was derived from a Lushootseed (a language spoken by several Native American tribes in the Puget Sound region of Washington) word (*gʷídəq*) meaning "dig deep." It is believed that the name has absolutely nothing to do with ducks or any other aquatic fowl but may be more the result of poor transcription than anything else. There have been a variety of spellings throughout history, including *goeduck*, *gweduc*, *goiduck*, and *gweduck*. In China, it is known simply as the "elephant-trunk clam" (easy to say, easy to remember).

These clams are further distinguished by the fact that they are the largest burrowing clams in the world. While their average weight is often between one and three pounds when harvested, specimens up to fifteen pounds each (imagine that!) have been regularly pulled from the substrate in both Washington and British Columbia.

FAST FACT: A single adult geoduck can filter about thirty gallons of water every day.

Geoducks are also considered "old timers" in the clam world (or in the overall animal world, for that matter). Several specimens exceeding 100 years of age have been uncovered, with the oldest recorded geoduck clam coming in at a whopping 168 years old. Part of their longevity may be due

to the lack of wear and tear in their quite muddy and quite sedate ecological niche.

Pacific Littleneck (*Leukoma staminea*)

Pacific littlenecks are a favorite steamer clam up and down the West Coast. These clams typically have thick, oval, and chalky-white to brownish shells. Their shells are also decorated with narrow, radiating ribs and concentric growth ridges. They also have short necks with siphons fused together. They burrow from four to ten inches into sandy, muddy, and sometimes gravely sediments. Littlenecks are fast-growing and fairly long-living clams, often reaching sixteen years of age. However, like other clams, their growth rate tends to slow down as they get older.

Pismo Clam (*Tivela stultorum*)

The Pismo clam is one of the most important edible bivalves along the California coast. It can be found along the entire coast south of Monterey Bay and down into Mexico. However, its primary territory (and greatest abundance)is in and around San Luis Obispo County in central California. Pismo clams are distinguished by being one of the largest clams found along the coast. If left alone, these clams will grow up to seven inches in length and ten years in age. However, hungry sea otters make sure very few clams ever reach this size (or age).

Pismo clams grow approximately three-quarters of an inch each year for at least the first five years. After that, they seldom grow more than an eighth of an inch per year. These clams are especially relished by clammers, as they can be found just under the sand's surface during periods of high tide.

> **FAST FACT:** A three-inch Pismo clam filters an average of 5,800 gallons of water per year. However, this quantity of water contains a mere 3.88 ounces of food. Thus, it can be said that these clams are very light eaters.

Butter Clam (*Saxidomus gigantea*)

Butter clams frequent the west coast of North America from the Aleutian Islands in Alaska down to Monterey Bay, California. The densest populations of this species are in Puget Sound and the Strait of Juan de Fuca in Washington. Slightly smaller than some of its other relatives, they average three inches in length. Their shell is usually light gray with white coloration and smooth ridges. Often located in the intertidal zone, butter clams bury themselves up to a depth of about one foot. They prefer a rough substrate composed of gravel and broken shells.

While butter clams are a frequent target of west coast clammers, they also has several scientific uses. They have frequently been used by archaeologists for research into ancient peoples and their lifestyles. Interestingly, butter clamshells up to 190,000 years old have been found in Lynn Point and Willapa Bay, Washington. And butter clams have been "employed" as "markers" in assessing the amount of pollution in various oceanic environments. An analysis of their shells can help researchers reconstruct and validate changes in sea surface temperatures over long periods of time.

> **FAST FACT:** Butter clams can store paralytic shellfish poisons (obtained from eating the dinoflagellate *Alexandrium catanella*) in their siphons for up to two years. These poisons tend to deter any potential predators (sea otters and sea birds) intent on nibbling on those siphons as they stick up out of the sand. Just as a precaution, human predators are always advised to clip off and discard the black ends of the siphons before ingesting these delicacies.

In the Pacific Northwest, they may be known by a variety of names, including Martha Washingtons, Washington clams, beefsteak clams, and quahogs. With all those different appellations, you might expect them to be a significant commercial and recreational clam . . . and indeed they are! In fact, there are legions of people who will claim that butter clams make the best chowder of all the varieties of bivalves. (New Englanders, take note.)

Dark Mahogany Clam (*Nuttallia obscurata*)

Dark mahogany clams are hitchhikers . . . or, at least they know how to take advantage of some free travel. You see, these clams are actually native to Japan and Korea. Up until 1992, they had never been found outside those two countries (they may also be found in Chinese waters). Then, they saw an opportunity for an extended trip and snuck into the ballast water of one or more ships headed eastward to the Pacific Northwest. British Columbia was on their itinerary first, and then they decided to venture southward to Washington and Oregon (perhaps searching for warmer waters). Slow travelers, they eventually made it down to Oregon sometime

around 2000 and took up residence in several estuaries along the Oregon coastline, where they have now established permanent residency.

This species has flat, oval shells that are slightly elongated on the rear end. They can be most easily recognized by their shiny dark brown periostracum—a distinctive feature that makes every clam look like it has been painted with spar varnish. To carry that colorful distinction even further, the inside of the shell is a vivid purple, thus making this clam easy to identify (or easy to differentiate) from every other clam up and down the Oregon coastline. Like many other species, dark mahogany clams have two separate siphons close together—clammers need only to locate the siphon holes (they are often buried in less than ten inches of sand) to find these most edible delicacies. Many folks prefer to steam or grill these delicious bivalves as a most delightful accompaniment to summertime meals.

Pacific Razor Clam (*Siliqua patula*)

You would think that an animal that has not one, but two, separate annual festivals* in its honor would be the talk of the town . . . or, at least, the talk of Washington. And you'd be right. This clam may be one of the most celebrated in the entire Northwest; it is certainly one of the most prized. Although most often associated with the state of Washington, they can be found in Alaska, Oregon, and down the California coast to Pismo Beach. During razor clam season, clammers travel long miles just to get their daily limit of fifteen clams each.

As you might suspect from their name, these bivalves have shells that are razor sharp. At maturity they may grow up to seven inches in length. For most folks, the preferred way of cooking them is to panfry them, because their thinness requires that they be cooked quickly. If cooked too long, they will remind you of the last time you chewed some old shoe leather (not that I'm criticizing your culinary predilections).

> **FAST FACT:** Adult razor clams have the ability to dig up to a foot a minute, and some have been recorded at depths of more than four feet. What makes this ability even more astounding is that they are incapable of horizontal movements; they can only move vertically through the sand.

---

* The Ocean Shores Razor Clam Festival in Ocean Shores, Washington (end of March), and the Razor Clam Festival in Long Beach, Washington (3rd weekend in April).

Juvenile razor clams begin their life in the first few inches of subtidal and intertidal regions of open coastline and various coastal bays. As they get larger, they will dig deeper into the sand. Besides being fast diggers, they are also fast growers. Most clams will reach a harvestable size of about three and a half inches by the end of their first year. If left undiscovered by human clammers, they will grow an additional inch in their second year. Most of that growth takes place during the spring and summer, when warm water and an abundant food supply predominate. Growth slows precipitously in the fall and winter. Growth also slows down sometime after the second year, when most of a clam's energy is devoted to reproduction rather than to growth.

Razor clams.

Pacific Gaper Clam (*Tresus nuttallii*)

Do any traveling along the highways and byways of America and you'll most likely come across the classic iconic feature of every road or intersection—the Great American Hitchhiker. Thumb up in the air, a look of abject passivity on his (or her) face, and an overwhelming need to be someplace else (for free) highlight these figures with cardboard signs and battered backpacks.

Human highway hitchhikers are only part of the pantheon of critters and beings hooking for a free ride through life. Humans willingly transport humans; and, truth be told, animals transport animals—perhaps not willingly, but perhaps more due to convenience than anything else.

So it is with the Pacific gaper clam, which provides transportation, protection, and comfort for certain types of pea crabs—most notably *Pinnixa faba*, *Pinnixa littoralis*, and *Fabia subquadrata*. Ecologists refer to this "partnership" as commensalism, a relationship between two organisms where one benefits without affecting the other.* The pea crabs secure a safe and hospitable habitat inside the clamshell, and the clam (presumably) is entirely unaffected and, who knows, perhaps no less the wiser.

This clam gets its name because its two shells do not close completely—they gape on the posterior end where the neck protrudes. Also known as horse clams (sorry, but I fail to see the resemblance), they often grow to lengths of eight inches and have been known to burrow to depths of three feet or more. The shell is, for the most part, creamy white; however, it may also be covered with a brown varnishlike coating. They are typically found in clusters in very sandy sediments.

Manila Clam (*Venerupis philippinarum*)

Look at the variety of clams that inhabit the west coast of the United States—particularly those native to the Orient—and you might get the impression that the travel agents in Japan, Korea, and China are working overtime to extol the virtues of western living to entire populations of far-eastern bivalves. For, once again, here is another creature that hitched a ride (in oyster seed shipments from Japan) from its ancestral home all the way across the Pacific, not just to visit but to eventually take up permanent residence in British Columbia and all three west-coast states. Not that we shouldn't appreciate all the visitors and their economic influence (tourists are always welcome); it's just that the US Immigration and Customs Enforcement may be taking their import just a little too lightly.

Manilas, though relatively recent immigrants (since the 1920s), have become a significant commercial blessing for Washington, accounting for at least 50 percent of all the commercial hard-shell landings in that state.

---

* Another example of commensalism is barnacles affixed to a whale. Examples of other ecological "partnerships" include mutualism, a relationship in which both organisms benefit (for example, bacteria living inside the intestines of cows); amensalism, a situation in which one organism is harmed while the other is left unaffected (cattle trampling grass); and parasitism, where one organism benefits while the other is harmed (fleas on dogs).

They are also embraced by legions of recreational clammers, since they are relatively shallow burrowers (often found in the first two inches of substrate) and easily dug out by young and old alike. They are especially prized as steamers (needing about three to five minutes to steam open) and are some of the sweetest clams on the market. They are frequently used in pastas and soups.

Manila clams are distinguished by their small size and attractive shells. The ridged shells often have deep, wide bars of color that make them easy to identify—particularly since they are often found alongside Pacific little-necks. In the wild they will live between seven and ten years and reach an overall length of approximately eight inches. Commercial clams are normally three to four years old with an overall shell length of about three inches.

Soft-Shell Clams (*Mya arenaria*)

M. arenaria has a long evolutionary history, having originated in the Pacific Ocean during the Miocene era (23,030,000 to 5,332,000 years ago). It slowly extended its range in the early Pliocene (5,332,000 to 2,588,000 years ago) to what is now the Atlantic Ocean, as well as around the European continent. For unknown reasons, the Pacific and European populations went extinct some time during the early Pleistocene (2,588,000 to 11,700 years ago), leaving only the northeast Atlantic population extant. It was that population, subsequently spread via humans, that reached its current distribution.

Soft-shell clams get their designation not because their shells are soft and mushy but rather because their shells are thinner than those of hard-shell clams. (They can be easily crushed when held by humans.) For the most part, soft-shell clams have thin, oval, and elongated shells. Quite often the shells are chalky white with a thin, brittle covering that varies in color from brownish to gray. Evolution has not been kind to these creatures—they cannot close their shells completely, consequently they gape open at both ends. It should be pointed out that the clam's body is often longer than its shell (another evolutionary boo-boo), allowing its foot, along with two long leathery siphons, to protrude from either end. While these clams will never be featured on *People* magazine's list of sexiest celebrities, they do have the distinction of having been around quite a bit longer than any self-absorbed actor or silicone-breasted starlet.

> **FAST FACT:** Soft-shell clams are a favorite of sea gulls. The gulls pull the clams from the sand, climb up to about twenty feet, and then drop the clam on a hard surface (e.g., an asphalt parking lot), breaking the shell. Then they swoop down quickly to eat the soft parts of the clam before others can get to it.

East Coast Soft-Shells

Steamed steamer clams.

I suppose if soft-shell clams had emotions, they would also have some major personality disorders. That's because these clams have about as many names as those who engage in "the world's oldest profession."* Travel up and down the east coast of the United States and you're likely to hear the following terms, all of which refer to the same exact animal: Ipswich clams,

---

* Not that you should look this up. I did the necessary research for you and discovered more than forty different terms for harlots, strumpets, trollops, courtesans, and ladies of assignation.

steamers, longnecks, maninose, nannynose, piss clams,* squirt clams, or Essex clams. Yup, it's an identity crisis in the making!

> **FAST FACT:** It's no secret among seafood suppliers and restaurants that most of the soft-shell clams currently sold as "Ipswich" clams, even in Ipswich, Massachusetts, in fact come from Maine.

Soft-shell clams can be found up and down the east coast from Canada south to Florida. Their preferred habitats include muddy tidal flats and estuaries. They will often burrow their way six to ten inches into soft sediments. They typically don't like to venture far and will spend most of their entire adult life in one place. Depending on location and the persistence of various predators, soft-shell clams can live for up to twelve years.

West Coast Soft-Shells

People on the west coast like things straight and simple. No complications, no misunderstandings, no "we'll give you one name if you live here, but another name if you live there." As a result, animals on the west coast have names that are also straight, simple, and down-to-earth. For example, folks in California, Oregon, Washington, and Alaska call soft-shell clams "soft-shells." Yup, it's that simple. No regional appellations, no geographical nomenclatures, and no monikers that say one thing and actually mean quite another (see "Ipswich clams" above).

Or, as the French would say, *très simple.*

Soft-shell clams (*Mya arenarea*) were originally thought to be naturalized imports from the east coast, possibly introduced in 1869 with shipments of oysters to San Francisco. Recent studies, however, indicate they may actually be native to western shores. Found in scattered groupings in intertidal sand and muddy areas, they live about four to twelve inches below the surface. Growing to a length of about four inches, their brittle shell is chalky-white, with a periostracum† around its margins. Western soft-shell clams are particularly fond of low-saline waters, so they are often prevalent where creeks and rivers flow into the ocean.

---

* They got this nickname from the fact that the siphon often sticks up through the sand and the clams unwittingly reveal their location by spurting water, much as you might do if a gargantuan creature (dressed in ratty clothing) with an oversized weapon (shovel) stepped near you.

† A thin organic skin that serves as the outermost layer of the shell.

These soft-shells are typically medium-size clams; and, like their eastern relatives, they have shells that are easily broken. Like their cousins, their shell is rounded at the foot end and rather pointed at the siphon end. The external surface of the shell is frequently marked with concentric rings. They will normally bury to a depth of eight to fourteen inches. All in all, they are a favorite target of clammers and are often taken with shovels or standard garden forks.

∞

That there is an infinite variety of clams to satisfy the hunting inclinations of any clammer or the culinary preferences of almost any diner is a given. However, the fact that so many of these bivalves are saddled with a plethora of names, appellations, designations, titles, eponyms, sobriquets, *nom de plumes*, and nicknames is sufficient to cause not only nomenclatural confusion but culinary befuddlement.

Indeed, it almost seems as though you need a lexicon whenever you go out to eat. Ordering steamers in Massachusetts may be quite different than ordering steamers in Texas, for example. Dining on soft-shells in New Hampshire could be considerably different that dining on soft-shells in Seattle. And, a quahog in Rhode Island may be quite different that a quahog consumed in the panhandle of Florida.

However, truth be told, clams have it pretty easy when it comes to names—specifically unusual or decidedly strange names. On the other hand, humans have the unique habit of giving our progeny names that no creature in the animal world would even think of using. Witness the following list of brand-new baby names as reported recently (2014) by the Social Security Administration (which means these individuals are stuck with these names . . . throughout their school days, their working life, and well into their retirement years):

Carrion (Isn't this something vultures eat?)
Harshit (This is actually a Sanskrit name, unfortunately one that lends itself to lots of playground teasing.)
Vadar (May the Force be with you!)
Goodness (Sounds like my breakfast cereal.)
Vegas (I wouldn't gamble on this one.)
Ikea (It's certain the bedroom furniture and the child will be compatible.)

Emperor (Could have been called "King of the Hill.")

Princeton (Don't apply to Harvard.)

Rage (Will probably be the leader of the local chapter of Hell's Angels.)

Shady (Will probably become the neighborhood "loan shark.")

Money (Sorry, but this one makes no cents to me.)

Chapter 6

Anatomy and Physiology

No man should marry until he has studied anatomy and dissected at least one woman.

—Honoré de Balzac

GROWING UP, I LOVED SCIENCE FICTION MOVIES—those cheaply made, cinematic dramas that featured radioactive creatures, horrible mutant life-forms from distant galaxies sent forth to take over the earth, or a (very irritated) prehistoric reptilian creature bent on destroying as much urban property as possible in seventy-five minutes or less. These were the movies I cherished—pseudo-scientific dramas that enabled a preadolescent boy to escape to another dimension every Saturday afternoon.

Now lovingly referred to as "B" movies, those classics of cinema took incredible biological liberties with the fauna of this world (and many other worlds as well). Some of the most memorable creatures of the era were undeniably gargantuan critters such as Godzilla, Rodan, The Beast from 20,000 Fathoms, and the creature in *20 Million Miles to Earth*, critters who haphazardly stomped cars, smashed buildings, and destroyed entire towns to the extreme delight of a movie theater filled with preteen boys (and, occasionally, their preteen dates).

One of the most intriguing science fiction films was *Attack of the Giant Leeches*, a 1959 movie produced by Roger Corman ("The King of B Movies"). Made on a miniscule budget (a Corman trademark), it took advantage of 1950s Cold War fears, specifically what would happen when some already pretty disgusting creatures were exposed to atomic radiation and made even more disgusting (read: considerably larger than the family car) as a result of some very rapid genetic mutations (and some hastily constructed latex costumes).

The entire movie was filmed in just eight days (another Corman trademark). Although the story is supposed to take place in the Florida Everglades, most of the outdoor sequences were shot at the Los Angeles County Arboretum and Botanic Garden within driving distance of where I grew up. But, for me, the movie was a definite "must-see" when I saw the promotional film poster with this tagline:

CRAWLING HORROR . . .
Rising from the depths of Hell . . .
To kill and conquer!

The fact that there were a couple of nubile vixens pictured on the poster may have also been further cinematic inducement to pedal down to the Aero Theater one particular Saturday afternoon to see this film.

FAST FACT: One of the central characters in the film, Liz Walker, was played by actress Yvette Vickers. Vickers had previously appeared as the centerfold in the July 1959 issue of *Playboy* magazine.

Since several marine creatures (such as leeches) were being irradiated in the 1950s—and characteristically transformed into larger-than-life monsters with larger-than-life appetites—I could never figure out why clams weren't considered as potential science fiction movie beasts. For example, I could just imagine an atomic bomb being tested in the southern California desert with a resultant cloud of radioactive matter drifting over a certain beach just south of Los Angeles. The next day a young couple, out for a romantic walk on that same beach, strolls around a rocky outcropping and sees several gargantuan clams (say, ten feet tall or so) opening and shutting their shells and, of course, sliding over the sand towards them. The young woman screams, the young man stands in front of her to ward off these apparently alien beings, and some cheap violin music begins playing rapidly in the background.

The movie (*Attack of the Killer Clams:* **BEACHFRONT HORROR!** Rising from the Sand…To conquer and consume!) is off and running, and

for the next sixty-five minutes or so the Navy and some local marine biologists argue about the best way to eradicate these very dangerous marine creatures. (Incidentally, the young couple is never heard from again.) Along the way, tourists from Montana, families with large umbrellas and small children, and old ladies walking their dogs on the beach are unceremoniously consumed by the molluscular monsters—just imagine the sound effects! Finally, someone contacts the Air Force, and they scramble two to three jets together to bomb the hell out of the gigantic bivalves (at this point, a General makes a bad joke about clam chowder) and the local surfers again take to the waves as the sun sets over the ocean (cue a Beach Boys song).

Admittedly, it's not much of a plot, but it wasn't the plot that was important in those films, it was the overwhelming size (and threat) of the creatures that drew in preadolescent moviegoers to theaters nationwide. Still, no one sought to employ gargantuan human-sucking clams as the villains in a "B" movie. I still can't figure out why.

Classification of Clams

At this point in our discussion of human-consuming clams it becomes necessary for us to digress and discuss an element of the biological sciences known as taxonomy. If you went back and asked some ancient linguists, they would tell you that the word *taxis* means "arrangement" or "order" and the word *nomen* means "name." Taxonomy is the practice and science of classification; in other words, how plants and animals are named according to their presumed natural relationships. For example, you and I have hair (although you, most likely, have more than I), we communicate via spoken language, and we walk on two feet. Sea squirts, on the other hand, have no hair, do not communicate through words, and cannot walk. Taxonomically speaking, humans belong in one group (or class) and sea squirts in another.

FAST FACT: As difficult as it may seem, sea squirts and humans do share some commonalities. We both belong to the phylum known as *Cordata*—animals that possess a notochord* and are bilaterally symmetrical (among other features). Fortunately, we each belong to two different classes within that phylum, *Ascidiacea* and *Mammalia*.

---

* A long flexible rod of cells that forms the supporting axis of the body. In higher vertebrates it eventually becomes part of the vertebral column.

Now, if we were to ask your friendly neighborhood taxonomist about hard clams, one of the subjects of our book, he or she would grab a taxonomic textbook and map out the taxonomy for our friends as detailed below.

Hard Clam Taxonomy

Taxonomy	Name	Meaning
Kingdom	*Animalia*	Animal
Phylum	*Mollusca*	Joint-legged animals
Class	*Bivalvia*	Marine and freshwater mollusks that have a shell in two hinged parts.
Order	*Veneroidae*	
Family	*Veneridae*	There are 53 genera and about 500 species.
Genus	*Mercenaria*	
Species		There are more than 15,000 different species of clams throughout the world—about 500 live in fresh water, the rest in salt water.

As you can see in the chart above, hard clams belong to a very distinctive phylum, the Mollusca. Mollusks are invertebrate animals with (in general) three body regions: a head, a visceral mass, and a "foot." The head contains the sense organs and "brain," while the visceral mass contains all the internal organs. The "foot" is the muscular lower part of the body that is typically in contact with the substrate (the surface on which an organism lives). Mollusks usually have a shell (although some do not).

The word mollusk is derived from the French *mollusque*, which originated from the Latin *molluscus*, from *mollis*, soft. *Molluscus* was itself an adaptation of Aristotle's τά μαλάκια, "the soft things." Included within this phylum are ten classes of animals, two of which are entirely extinct. Here is how those classes would look in chart form (with a few examples of each):

Mollusks

Class	Organisms	Living Species
Caudofoveata	wormlike organisms	120
Solenogastres	wormlike organisms	200
Polyplacophora	chitons	1,000
Monoplacophora	An ancient lineage of mollusks with caplike shells	31
Gastropoda	All the snails and slugs, including abalone, limpets, conch, nudibranchs, sea hares, sea butterfly	70,000
Cephalopoda	squid, octopus, cuttlefish, nautilus	900
Bivalvia	clams, oysters, scallops, geoducks, mussels	20,000
Scaphopoda	tusk shells	500
Rostroconchia	fossils; probable ancestors of bivalves	extinct
Helcionelloida	fossils; snaillike organisms	extinct

The human body is a pretty complex assembly of various parts (some a little chunkier than others, at least for certain authors) all working together to get you through the day and maintain your overall well-being. Truth be told, your body is a fabulously complete "anatomical machine" of eleven interlocking and interrelated systems.*

In essence, you're a pretty complex character. For clams, not so much.

You see, clams are fairly simple critters, physiologically speaking. In general, clams have two shells, two siphons used to draw in or expel sea water, a single hatchet-shaped foot† used to burrow into sand or mud, and a whole bunch of really delicious "innards."

Pretty simple.

---

* Your anatomical systems are: skeletal, muscular, cardiovascular, digestive, endocrine, nervous, respiratory, immune/lymphatic, urinary, reproductive, and integumentary (skin, hair, nails, and exocrine glands).

† You're undoubtedly familiar with other one-footed animals, those that tend to frequent vegetable and flower gardens: snails and slugs.

Now, let's pluck a clam from the waters of a convenient eastern shore-line and take a look at it. Let's call our specimen "Chris." The name "Chris" was chosen for a simple reason—most clams have the capacity to change their sex. They often start off as males, but later in life, may "decide" to change their gender to female. Thus, at different times in their life they are capable of producing both male and female gametes. (This sexual conundrum is explored in greater detail in Chapter 7. Suffice it to say, the concept of "boy meets girl" has a completely different connotation for clams.)

"Chris" is, therefore, a name that can be used for either male or female creatures, such as humans; or, in this case, our resident clam.

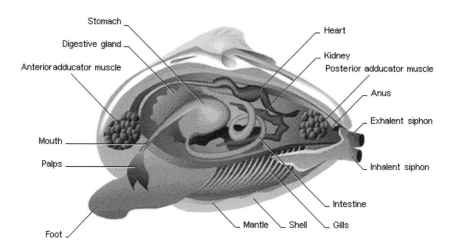

A general clam diagram.

As you'll recall from the earlier chart, clams are part of a class of animals known as Bivalvia. Other class members include oysters, mussels, scallops, and geoducks. The name Bivalvia was first used in 1758 to describe the shell, which is composed of two valves. The term bivalve is derived from the Latin *bis*, meaning "two", and *valvae*, meaning "leaves of a door." Bivalves, therefore, are creatures with a shell consisting of two asymmetrically

rounded halves that are mirror images of each other. The two halves are joined at one edge by a flexible ligament called the "hinge". To make it simple, you might want to think of clams as soft-bodied creatures encased in calcium "suitcases."

In the United States we use the word "bivalve" to identify three very broad categories of animals. These include those critters we know as clams—bivalves that burrow in sediment and that we consume at hundreds of seafood restaurant and clam shacks around the country. A second broad group includes bivalves that attach themselves to underwater surfaces, such as oysters and mussels. The third category is those bivalves that swim, and are often migratory, such as scallops.

> **FAST FACT:** Scallops have up to one hundred simple eyes arranged around the edges of their two mantles. Each eye contains two retina types—one that can detect light, the other darkness (such as the shadow of an approaching predator). Interestingly, all scallop eyes are blue.

Skeletal System

But, let's get back to "Chris." At first glance, you would probably say that Chris (at least on the outside) looks relatively undistinguished. It is just those two bivalve shells, each a mirror image of the other. Whether you look at Chris from the top or from the bottom, it looks the same. One of the distinguishing characteristics about clams is that they, along with a few other mollusks, always have two shells. Some mollusks, such as various species of snails, also have two shells, but they start their larval existence with only one. Chris, on the other hand, developed two shells early in life, will live with two shells throughout its life, and will die with two shells.

Chris's two shells are composed of a combination of calcite (a common mineral, often dissolved in seawater, and a constituent of sedimentary rocks such as limestone) and aragonite (a naturally forming substance in almost all mollusk shells). This composition ensures that Chris will have a form of environmental and predatory protection, much like an underwater suit of armor. The shells make it very difficult (though not impossible) for other marine creatures to get through that hardened material.*

---

* Whelks (found from Cape Cod south to Florida) are some pretty dangerous predators for clams. To begin, a whelk uses its foot to grasp and position a clam. Then, the whelk

While we may be most familiar with clamshells that are slightly oval in shape, Chris and its relatives can have a wide variety of shell shapes. Some clamshells may be oval, others globular, some rectangular, while a few species have elongated shells to aid in burrowing. As you might expect, the shape is species-specific and has evolved over time to enhance protection and facilitate feeding.

Chris, like all clams, will only open its shells for two specific reasons: in order to feed and in order to reproduce. At all other times, Chris's shells are closed as protection against carnivorous sea creatures (like sea stars) that would love to dine on the soft insides. However, irrespective of the size or shape of the shell, it's the insides that are identical in all species of clams, including Chris.

In humans, the skeletal system provides support and protection for the soft tissues that make up the rest of your body. It also includes such things as tendons, ligaments, and cartilage. Our skeletal system provides several vital functions—support, protection, movement, endocrine regulation, calcium storage, and blood cell production.

FAST FACT: Human babies are born with anywhere from 300 to 350 bones. Several of those bones fuse together as the body grows (such as the bones in the cranium). By the time a child reaches the age of nine, his/her body will have just 206 individual bones.

Some of the critical elements of your skeletal system are your ligaments. Like muscles, ligaments are connective tissues. However, unlike muscles, ligaments are designed to connect bone to bone; they typically serve to hold structures together and keep them stable. That connection is known as a joint. For example, we know there are seven different types of ligaments

---

inserts the sharp edge of *its* shell between the clam's two shells. Using its foot and body, the whelk gently rocks the clam back and forth. In the process, the whelk's shell edge slowly chips away at the outer margin of the clam's shell. Over time, the clam is slowly and gradually forced open as the whelk continues to apply pressure. Eventually, the sharp edge of the whelk's shell cuts into the soft tissue of the clam. Then, the feast begins: the whelk inserts its proboscis inside the clam and consumes the inhabitant. Being eaten inside your own house is not a pleasant way to go. (But it would make for one hell of a movie on the Syfy Channel, don't you think?)

that are working to keep your knee functional. When one or more of those ligaments is damaged or diseased, then it may be necessary to repair the knee to make it functional (and pain free) again. That is to say, we want to return the knee to a state of stability again.

Chris, too, has ligaments—two, actually. Its shells (rather than bones) are held together at the top by those two ligaments. Depending on the species, those ligaments will either be under compression when the clam has its shells closed or they will be stretched when the clam has its shells open. In each case, relaxation of powerful muscles (known as the posterior and anterior adductor muscles) used to close the shell causes the shells to open because of relaxation of the stress on the hinge ligaments. Consequently, the clam's shells close because of its active muscle power, and open when the muscles relax or when the clam dies.

If you've ever tried to open a living clam with your bare hands, you know how difficult it is to pry the shells apart. Clams have the ability to hold their shells tightly closed with their previously described very powerful muscles. Those muscles, attached to the inner surface of both shells, constantly regulate the closing of the two shells. However, there are times when those muscles are put to the test—those times when extremely determined critters, such as sea stars (and the aforementioned whelks), desire a delicious clam dinner.

Inappropriately referred to as starfish (they are not fish), sea stars can cause havoc within a colony of clams. (Truth be told, if clams could experience fear, sea stars would probably be some of the most feared predators throughout the clam world.) While these creatures may be beautiful to behold (at least to us humans), they are decidedly deadly to clams.

You see, sea stars eat clams in a most unusual way. First, they position themselves over the shell of a convenient clam (Not "Chris," we hope!). Gripping a nearby rock with some of its suckers to anchor itself, the sea star will hold on to both halves of the clam's shell with several of its other suckers. Then the sea star begins to pull the two clamshell halves apart. It's a classic case of tug-of-war; however, the final results are inevitably one sided. Despite the clam's strong muscles, the sea star is doggedly persistent. In fact, a sea star may pull for several hours or even several days, eventually separating the clam's shells just a little. After a sea star separates the clam's shells, it pushes its stomach into the clam, inserting it inside out. The sea star then secretes digestive juices into the unlucky mollusk and digests it right inside its own shells. In the end, the sea star's patience is rewarded by a most-favored meal.

> **FAST FACT:** One of the most unusual marine creatures is the *Linckia* sea star. This animal is able to pull itself in separate directions until it breaks into two parts. Each of the two parts can then grow into a new animal.

Brain Function

In the first chapter of this book, as you may recall, we spent time talking about some of the great thinkers of our time—folks like Thomas Jefferson, Albert Einstein, and, of course, our friend Leonardo da Vinci. While you or I may never be included in the company of those thoughtful individuals, we all have one thing in common—a brain. While it is not my intent to get into an extended discussion of this three-pound collection of neurons, glial glands, and blood vessels, suffice it to say that this organ is the font of our intelligence, the interpreter of the world around us, the source of our body movements, and the explanation for our behaviors. Or, as one writer put it, "The brain is the crown jewel of the human body."[*]

Now, just for a moment, try to imagine a creature without a brain. How would that creature be able to function? How would it control its eyes, its mouth, its motor functions, and its respiration? How would it survive, procreate, defend itself from its enemies? As you might guess, the absence of a brain places severe limitations on one's ability to get through life, not to mention just trying to get a simple meal.

While clams are not endowed with a brain (after all, who needs a brain if you basically sit in one place through your life and filter water?), they have what is often referred to as a "simplified brain," known to biologists as *ganglia*. While most creatures have a single brain, many species of clams have two coordinated ganglia. One, known as the pedal ganglia, is responsible for controlling the clam's foot. The other, known as the visceral ganglia, regulates the clam's internal organs.

Ask a zoologist for a definition of ganglia, and he or she will likely tell you that it is a biological tissue mass, or a mass of nerve cell bodies. Human brains, such as yours and mine, are capable of thought and reason; however, ganglia are only capable of controlling the most rudimentary of biological functions: moving the foot, for example, or moving food along the digestive tract. In short, clams cannot think, they can only do. It's a classic case of stimulus/response—no pondering, no thought, and no

---

[*] "Brain Basics: Know Your Brain." *National Institute of Neurological Disorders and Stroke*.

contemplation. Our friend Chris is essentially a creature without reason, an organism without a conscience.

Although clams do not possess thinking organs, they have the capacity to sense their surroundings. For example, many species of clams have a series of tentacles with chemoreceptor cells used to taste the surrounding water. By tasting the water, Chris can determine if food is nearby or if a fellow clam is in the immediate area (always good to know when the possibility of reproduction is in the wind, so to speak).

Another notable sensory organ is the clam's osphradium, a patch of sensory cells located below the posterior adductor muscle (the muscle that opens and closes Chris's two shells). Chris sometimes uses this organ to taste the water or measure its turbidity—the cloudiness or haziness of water due to particulate matter (dirt) suspended in the water.

Chris and its relatives also have an additional sensory feature, a collection of organs known as statocysts. These organs help the clam sense and correct its orientation, that is, is it right side up or upside down? And, if it is right side up, is it tilted to one side or the other?* This orientation is critical in helping ensure that the clam is able to feed efficiently.

Respiration

The human respiratory system is responsible for the intake and exchange of oxygen and carbon dioxide. Since we are terrestrial animals, our respiration takes places in organs known as lungs. For marine creatures, their respiration is considerably different.

If you look closely at the earlier illustration, you'll notice that the generic mollusk pictured has a set of exceptionally large gills. The typical gills of mollusks are rod-shaped organs with thin filaments extending down off either side. Blood is pumped though the gills internally by a muscular heart, and water is moved over them externally by beating, hairlike projections called "cilia." *Cilium* (the plural is *cilia*) is Latin for "eyelash." In humans, cilia are found in the lining of the trachea, where they sweep mucus and dirt out of the lungs. In females, the beating of the cilia in the fallopian tubes moves the ovum from the ovary to the uterus.

Of even more interest is the fact that the clam's gills are covered in mucus. This mucus layer permits the clam to collect particles from the water that get trapped on its gills. The mucus moves because of the action

---

* I would imagine you could make a ton of money if you could just figure out a way to produce and package statocysts for the "regulars" at your local tavern.

of the cilia covering the clam's gills and helps flush these particles off its gills, either for use by the clam or for disposal.

In most bivalves, the gill filaments have become highly elaborated and enlarged—far larger than is necessary for respiration. Adjacent filaments become attached to one another, with small holes in the attachments, and in effect the clam's gills turn into a large net. Without thinking (remember, they have no brains), the clam keeps the surface of its gills constantly bathed in the aforementioned mucus. This helps the clam collect particles that "stick" to the gills, but now instead of moving the particles off the gills to clean them, the particles are moved to food grooves and conveyed to the mouth in a constantly flowing stream of mucus propelled by underlying cilia (try to imagine a department store escalator delivering scores of shoppers to the upper floors of the building continuously throughout the day).

Just for a moment, let's revisit Chris's shells. You see, the shells were originally secreted by a special internal tissue, called the mantle. The mantle extends down from the dorsal surface* over the entire animal on either side. In the area where the shells meet, the bottom edges of the mantle come together to form a curtain that can enclose a cavity between the shells. This cavity contains Chris's body and the expanded filtering gills. In several species of clams, this tissue curtain also is fused together to form a siphon.

The siphons are of two types: one is a tube for conveying water and food to the clam; the other one expels water and feces away from the clam. The two siphons are sometimes fused (depending on the species). Some years ago a researcher interested in discovering why clams are so successful ecologically determined that siphons evolved at least seven different times, so there are seven basic types of clam siphons. Interestingly, in all cases, the siphonal position on a clam marks the posterior surface of the animal.

FAST FACT: The paired siphons of the geoduck clam can be over thirty-nine inches long. (That's longer than your arm.)

---

* The top side of an animal is the dorsal surface; the bottom side of the animal is the ventral surface. For example, the dorsal fin of a shark is the fin on top—the fin that, moving through the water, causes elevated levels of fear, anxiety, and panic in any nearby swimmers (including a certain author who, in his youth, was approached by one [O. K., it was about two football fields away] during what was, up to that point, a particularly good day of body surfing).

If you recall any of your high school biology, you'll remember that your alimentary canal is the tube into which you stuff food (through your mouth) and which processes that food for use by your body. I like to think of my alimentary canal as a long tube (like underground subway tunnels in large metropolitan areas) with several interesting stops (or stations) along the way: the oral cavity, pharynx (throat), esophagus, stomach, small intestine, and large intestine. However, what we frequently forget is that our digestive system is designed to accomplish six basic processes, all of which enable us to obtain the energy and nutritional benefits of last night's steak dinner or fettuccine Alfredo: ingestion, secretion, mixing and movement, digestion, absorption, and excretion.

Like you, Chris also has a digestive system—albeit one considerably simpler than yours. To feed, Chris extends one siphon (the inhalant siphon) up into the water to suck water into the shell while remaining safely buried out of sight. Since clams are filter feeders, Chris uses the inhalant siphon to strain microscopic plants (algae) from the water column. A few of Chris's relatives use their inhalant siphons to vacuum the sediment around their burrow in order to collect necessary organic debris. This organic debris is generally rich in bacteria, which in turn are rich in nitrogen, and thus a good protein source. However, I would be remiss if I didn't mention that those animals that eat deposits or mud don't actually digest the mineral components of mud, but rather eat the bacteria and other microorganisms living in the mud. (Just like you might pick out the chips in a chocolate-chip cookie without eating the entire cookie.)

Chris feeds by filtering phytoplankton, bacteria, detritus, and dissolved organic matter from seawater through one its two siphons—both of which developed during the seed-clam stage. The incurrent siphon (which usually does not appear until a shell length of approximately one-sixteenth inch is reached) brings in nutrients and has tiny tentacles along its rim that sort out possible food particles. The tentacles act as the on/off switch switch for the incurrent siphon. When plenty of nutrients are suspended in the current, the tentacles tell Chris to go into the feeding mode; if the seawater is clouded with suspended particles of sand, mud, and other large debris, the tentacles shut the system down.

Seawater is pumped into the clam by its gills, and food particles are attached to the gills in that thick mucus solution. Cilia on the gills move the food particles slowly toward the clam's mouthparts, where tissues that line the entrance further sort out the food particles, chemically determining which will be ingested and which will be rejected. Rejected particles, called pseudofeces,* are moved by cilia to the base of Chris's other siphon. When a sufficient amount of waste has been collected, it is forcefully ejected by that siphon (known as the exhalent siphon) in a stream of water.

Circulatory System

The human heart is part of a complex system known as the cardiovascular system. This system includes not only the heart but blood vessels and the approximately ten and a half pints of blood that continuously move through your body.

> **FAST FACT:** Your heart beats approximately 101,000 times a day. During your lifetime it will beat about 3 billion times and pump about 800 million pints of blood or about 1 million barrels. That's enough to fill more than three supertankers.

Bivalves also have a circulatory system, albeit one completely different from yours. Clams have what is known as an open circulatory system—one that bathes the organs in hemolymph. Hemolymph is a fluid that fills the interior of an animal's body (notably invertebrates). It differs from blood in that it contains a copper-based protein that turns blue when exposed to oxygen. Vertebrate blood, on the other hand, is iron-based, giving it a characteristic red color.

Like other clams, Chris's heart has three chambers: two auricles receiving blood from the gills, and a single ventricle. By contrast, your heart has four chambers—two ventricles and two atriums. Chris's ventricle, on the other hand, is muscular and pumps hemolymph into the aorta, and through this to the rest of Chris's body. Oxygen is absorbed into the hemolymph in the gills, which hang down into the mantle cavity. The wall of the mantle cavity is a secondary respiratory surface and is well supplied with

---

* If you remember any of your high school Latin, you'll be able to decipher this word as "fake shit." Sufficiently intrigued by the definition, I did what any high school sophomore would do—I researched the term on Google. Needless to say, I came away with a plethora of documentation and information sufficient to build a whole new file (should I decide to write a sequel to this book—say *The Secret Life of Pseudofeces.*). If nothing else, I will have achieved the respect and admiration of my colleagues in academia. Now when I pass through the hallowed halls, they may well say, "Hey, did you hear about Tony?' "No, what?" "Yeah, it looks like he finally got his pseudofeces together!"

capillaries. Some species of clams, however, have no gills, with the mantle cavity being the only location for gas exchange. Other bivalves (such as mussels) can often survive for several hours out of water by closing their shells and keeping the mantle cavity filled with water. That's why they are frequent denizens of the intertidal areas of seashores around the world.

Mobility

You and I and every other human on the planet are bipedal. We use our feet for movement from one place to another, for kicking the tires in an auto-mobile showroom, for showing off our dancing skills on a cruise through the Caribbean, and for stomping our foot when our teenage son stays out much later than he should with the family car.

Chris, however, is a little different—clams only have one foot. A clam's foot is not at the end of its body, but rather located between the gills. In addition, a clam can do something with its single foot that we cannot do with our two feet—extend the foot in and out of its body. However, just like our two feet, a clam's single foot can be used to travel from place to place (a movement, I suspect, that would be similar to a three-legged race at the annual company picnic). If it wanted to (although it doesn't have the capacity to think about it), a clam (depending on its species) could move exceptionally fast.

Nevertheless, Chris, like most its relatives, generally moves very little. For the most part, clams are relatively sluggish or downright sedentary. Several species of clams (notably the ones we consume) burrow into the mud or sand and stay right where they are . . . like, forever. Since the food comes directly to a clam's mouth through water currents (the original "home delivery"), there is very little need for it to move around.

Some of Chris's relatives live on rocks or on coral substrates rather than in mud or sand. These critters fasten themselves to the substrate by the means of secretions from a secretory area called the "byssal gland" in the bottom center of the foot. This gland secretes liquid adhesive chemicals that harden on contact with water. In most cases, the clam places the byssal gland opening on the substrate and secretes some of the glue. The foot is then rapidly pulled away from the substrate and the glue hardens into a thread. The clam holds on to its end of the thread with a strong set of muscles, and the secretion sequence can occur again and again, so that finally the clam is held in place by mass of byssal threads.

Byssal attachment is remarkably durable and very strong. Consider that mussels in the intertidal zone can withstand the constant and often brutal pounding of surf without being detached. On shallow coral reef flats, the attachment of tropical clams to the substrate is by a byssal attachment and is capable of withstanding the unyielding pounding and pummeling of intense surf. Yet, as strong as they are, when necessary, clams can release their byssal threads and proceed to move about.

Clams may appear to be fairly simple creatures. However, they are also models of anatomical and physiological efficiency. The fact that we enjoy consuming them notwithstanding, their internal makeup is an evolutionary marvel. Still, I can't figure out why no one demonized clams in a sci-fi flick. How come nobody saw the obvious advantages of casting clams as some really bizarre life form from another world that devour hedge fund managers and other nonessential life forms on a violent rampage through some urban jungle? After all, clams certainly look and act like aliens with their symmetrical shells, their odd feeding habits, their elongated siphons, their liquid adhesive chemicals, and their overwhelming (yet ever-silent) presence on the shorelines of major industrialized countries.

Like enormous leeches and gargantuan prehistoric reptiles, they are creatures primed for cinematic glory. Wouldn't you agree?

Chapter 7

Sex Among the Bivalves

Sex appeal is fifty percent what you've got and fifty percent what people think you've got.

—Sophia Loren

INVITE ME TO YOUR NEXT COCKTAIL PARTY, AND I can assure you I will not discuss bovine insemination (even though I live in a semirural area with lots of farms in south-central Pennsylvania). It's not that I don't appreciate the efforts of humans to artificially inseminate female cows so as to create more cows and thus ensure the vitality of the dairy industry. It's just that I'm not yet convinced why humans have to get personally involved in a basic biological function that, at least as far as I know, cows have been handling pretty well on their own for a couple of thousand years. In short, why do we need to stick our nose (or, more specifically, our hands) in their business?

It wasn't until I read Mary Roach's bestselling book *Bonk: the Curious Coupling of Science and Sex* (2008) that I realized the scientific study of animal sex was a legitimate arm of biological investigation. It was Roach who alerted me to the fact that people can actually get huge amounts of grant money to observe something I can see on the farms I drive by on my way to work every day.

Roach also clued me in to a remarkably unknown pioneer in the field of animal sexuality—one Albert R. Shadle (1885–1963) who, at one time, was known, according to Roach, as "the world's foremost expert on the

117

sexuality of small woodland creatures" (a title that would make for some very interesting, and highly collectable, business cards). While Professor Shadle may have been an outstanding (nay, the only) expert in what is apparently a little known field of scientific investigation, it's still not clear to me how he was able to justify (to his college's department dean) an application for university funds for a study of "skunk and raccoon copulation and post-coitus behavior reactions."

After reading *Bonk* I began to wonder—is the study of reproductive habits of various creatures a legitimate scientific endeavor or a topic just on the edge of academic research? Is it practiced by serious scientists or just by overly hormonal undergraduates on spring break in Daytona Beach? Is it genuine or is it fringe?

I obviously needed some intellectual satisfaction! Thus I turned to the source of all knowledge—Google. I searched "animal sex" and was amazed to turn up more than 308 million sites (yes, 308,000,000) that advertised (among other things) the "Top Ten Best Animal Sex Videos," a museum in Germany with a "risqué animal sex exhibition," and, of course, "amateur animal sex" (suggesting that there must be a website or two devoted to "professional animal sex"). I reconsidered my phraseology and typed in "Animal reproductive habits" (only 9,690,000 websites popped up here). I was immediately presented with the following: "Squids may not seem like the sexiest animals around, but as it turns out, they . . . ."

It soon became evident that I was about to gain an unintentional membership into that legion of Internet voyeurs who spend many of their waking hours surfing all manner of sexually oriented websites. But to provide you, dear reader, with a complete and thorough picture of the lives and times of clams, I needed to delve into their reproductive habits. So (again, in the name of science), I returned to Google and typed in "clam reproduction" and turned up a mere 718,000 entries. Clearly, there isn't as much interest in the mating rituals of siphon-sucking, mud-inhabiting mollusks as there might be in those of, say, those sexy squids. However, I did discover that the topic has not gone entirely unnoticed.

Linnaeus and Nomenclature

The Swede Carl Linnaeus (1707–1778) was both a botanist and zoologist, and it was he who laid the foundations for the modern biological system of naming plants and animals. This system, now known as "binomial nomenclature," gives each biological specimen a name, specifically in

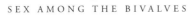

Latin,* composed of two parts. The first more general part of any scientific name is the name of the genus to which an organism belongs. A genus is a group of one or more closely related species thought to have evolved from a common ancestor and sharing many unique characters. The second part of an organism's scientific name is its species name. A species is a group of organisms that breed naturally with one another and do not breed with other such groups.

Here are a few familiar examples:

Human—*Homo sapiens*
House sparrow—*Passer domesticus*
Daisy—*Bellis perennis*
Lion—*Panthera leo*
Cat—*Felis catus*
Saguaro cactus—*Carnegiea gigantea*
Quahog clam—*Mercenaria mercenaria*

During his lifetime Linnaeus published several treatises outlining his taxonomic system and applying it to both botanical and zoological species. These publications included *Philosophia Botanica*, *Genera Plantarum*, *Species Plantarum*, and perhaps his most well-known publication, *Systema Naturae*. The first edition of *Systema Naturae* was twelve pages long. By the time it reached its tenth edition in 1758, it classified 4,400 species of animals and 7,700 species of plants. Linnaeus's genius lay in the fact that his book systematically reduced the unwieldy and ponderous names popular at the time into concise two-term (and occasionally three-term) names easily recognized throughout the scientific community and around the world.

Linnaeus was at the forefront of scientific inquiry—a man who transformed science (indeed, organized it) and who is remembered as both influential and memorable. German writer Johann Wolfgang von Goethe once wrote about Linnaeus, "With the exception of Shakespeare and Spinoza, I know no one among the no longer living who has influenced me more strongly." However influential he may have been, it seems as though

---

* Latin is often referred to as a "dead language." Language death—also known as language extinction—is a condition where people stop using a language (in favor of another language) or when the people who speak a language eventually die out. Interestingly, well over 6,000 languages have "died" since humans first began speaking.

Linnaeus occasionally strayed into areas not entirely taxonomic but more ribald in nature.

Consider his interpretation of clams.

In 1771, long after his success with *Systema Naturae*, Linnaeus published a treatise entitled *Fundamenta testaceologiae* (most of his scientific publications were written in Latin)—a description and examination of mollusks, specifically clams. He begins the publication discussing his classification of mollusks by their shells. He takes a few side trips to look at snails, scaphopods, and even a few tube worms. Eventually he settles on Conchae (clams, as well as chitons and barnacles).

Well into his treatise, and quite unexpectantly (considering his extremely conservative philosophy), he pens a most remarkable section, one seemingly out of place with taxonomy. This section offers a listing of technical terms for various clamshell parts. In describing the clam's hinge (between its two shells), he writes, "the notable protuberances above the hinge are called buttocks." He then proceeds to name other clam parts as though he was describing the sexual anatomy of human females. He describes a clam's *hymen* (the ligament connecting the two valves), *vulva, labia, mons veneris, umbos* (buttocks), and *anus*. He supplements these descriptions with a very detailed illustration of a common clamshell (which I am reluctant to provide you, dear reader, for fear of arousing any prurient interests among the audience).

These references seem out of character with Linnaeus's scientific protocol. They do, nevertheless, reflect a frequent sexual focus of so many essays of the time. In descriptions of flowers, many botanists often made reference to the male and female organs—stamens and pistils—as instruments of fertilization and reproduction. Later (highly respected) scientific authors, according to the aforementioned Mary Roach, also dabbled in explicit or implicit sexual innuendo in their treatises. Suffice it to say, the authorial tango of sex and science was not uncommon, it was just that it was unexpected, unconventional, and often considered unprofessional when practiced by stalwarts of the scientific elite.

Let's just say that Linnaeus, the contemporary luminary that he was, took quite a bit of professional flak for the terms he used in describing clamshells. (One English naturalist later referred to the terminology as "Linnaean obscenity.") If nothing else, Linnaeus's sexual innuendos did stimulate others well outside the scientific community to read this manifesto on clams. Now, Latin was sexy. Who knew?

Clam Mating Habits

In reality, clams have a distinguished method of mating.* To illustrate, imagine a wild and raucous celebration—say, after a professional football team has won the Super Bowl. People are swarming all over the streets of the city, horns are honking, sirens are blaring, folks are cheering, bells are ringing, strangers are hugging strangers, fireworks are being set off on every street corner, confetti is being thrown from office buildings, barrels of high-octane spirits are being quaffed, and festivities are at a fever pitch for hours on end. The city is in a state of pandemonium. "Our team won! Our team won! Our team won!" is shouted from every street corner, apartment building, and passing car. It is an unbelievable scene, one in which people completely let their hair down, lose all their inhibitions, and "party hearty" until the wee hours of the morning (and beyond) in an uninhibited demonstration and nonstop celebration.

That's what clam sex is like.

The animal world has several examples of unusual and distinctive reproductive practices (and consequences). Here are just a few of the most noteworthy:

- Male praying mantises literally put their life on the line when mating. If they happen to mate with a hungry female, the female will quite often turn her head around and bite off the male's head right in the middle of coitus (which, if you're wondering, continues even after the male has lost his mind).
- When a male honeybee mates with the queen bee, his genitals break away from his body and lodge themselves inside the queen's reproductive tract. As you might expect, this prevents other bees from mating with the queen. It also prevents the successful suitor from a long life—he quickly dies.

---

* One of nature's strangest reproductive acts involves whiptail lizards. You see, all whiptail lizards (a common desert reptile) are females; there are no males. So, how do they reproduce? Two females engage in pseudocopulation—one taking on the role of a male. Apparently, this stimulates egg production in both females. As a result, all whiptail lizards are clones of their mothers. Kinda makes dating websites completely irrelevant for these creatures.

- Male argonauts (animals similar to a squid) have a specialized tentacle that holds a small ball of spermatozoa at its end. When a male approaches a female, this tentacle detaches from the male and swims towards the prospective mate.
- When a male anglerfish is born, it immediately seeks out a much larger female. It then bites the female and releases a special enzyme that dissolves part of her flesh. This enables the male to fuse himself directly onto the female's body. Over time, the male eventually changes into a small bump on his partner. When the female is ready to reproduce, she obtains the necessary sperm from this bump.
- Male antechinuses (mouselike creatures) are known as nonstop sexual machines. They have been reported to engage in sexual intercourse with a single female for hours at a time. After they are done with one partner, they move on to another female for an additional (and protracted) sexual encounter—again . . . and again . . . and again! Quite often, males die simply from a lack of food and rest.
- Common bedbugs have a reproductive habit that may give you pause. It seems these critters mate through a process known as "traumatic insemination." That is, instead of the male inserting his sex organ into the female's reproductive tract, as is typical in most forms of animal copulation, instead he pierces the female with his hypodermic genitalia and ejaculates directly into the female's body cavity (ouch!).

To understand how clams mate, we need to understand a little biology. If you were to just look at a clam, you would have a difficult time identifying or even locating any discernable sexual organs—remember our friend Chris? That, and the fact that they spend their entire lives cooped up between two shells that are tightly closed, would seem to suggest that any kind of sexual activity would be, at best, challenging.

Then there's the matter of sexual identity. Normally, it's often quite difficult to know which clams are the girl clams and which are the boy clams (you could, of course, open the shell and check; but that would unnecessarily kill the clam). To make this sexual identity thing even more confusing, consider that some clams (predominantly quahogs) are protandrous hermaphrodites. That is, they begin life as males, but often change to females. About half the population will undergo this sex change, usually by the end of their first year.

Then, there is the other matter that some species of clams are "regular" hermaphrodites. That is, they have both male and female sexual organs and produce both male gametes (sperm) and female gametes (eggs). Although it is usually not possible for an individual to fertilize itself, it is possible for several nearby hermaphrodites to fertilize each other in a kind of neighborhood block party. ("I'll share my male gametes with you, if you'll share your female gametes with me.")*

Now, let's add in another factor. That is, in several species, fertilization occurs when there is a whole bunch of younger males and a whole bunch of older females. If all the clams in a certain area are the same age, then it is unlikely that any kind of sexual reproduction will occur. For those species of clams, the combination of younger males and older female is a big plus in the mating ritual. Since clams have been doing this for about 510 million years, maybe we should refer to them as the original "cougars."

As you can see, being a clam is not for the faint of heart . . . sexually speaking.

Now, let's take a look at how clams go about producing the necessary gametes that enables fertilization to happen in the first place. Here's a handy sequence:

Resting Stage
In this stage the clam completely (or almost completely) lacks gametes (eggs and sperm).

↘

Early Development Stage
The follicle walls thicken and immature gametes develop.

↘

Late Development Stage
The follicles rapidly expand to accommodate the larger and more numerous gametes.

---

* Consider snails, which are also hermaphrodites. In a bit of evolutionary whimsy, Mother Nature placed their male/female genitals on their necks, right behind their eye stalks. I'm not certain, but perhaps this is how the term "necking" originated.

↘

Ripe Stage
The follicles are fully expanded and thin walled. The lumen (the inside space of a tubular structure) of female follicles contains mature ova (eggs), while mature sperm dominate the lumen of male follicles. The germinal ducts have begun to expand and may contain a few mature gametes. Mature gametes are released and fertilization takes place externally.

↘

Resting (or Spent) Stage
The clam is completely devoid of gametes.

Mating Season

But, let's get one thing straight: clams don't release their gametes willy-nilly into the water whenever it's time to mate. There are two critical factors that come into play in order for the act of reproduction to take place. First of all, sexual reproduction among clams is highly dependent upon water temperature. Too cold, and the clams won't release their gametes; too warm, and any gametes that are released would, most likely, be killed because of extreme water temperatures.

That said, for fertilization to be successful, the water should be moderately warm; an optimal temperature is somewhere between seventy-three and seventy-nine degrees Fahrenheit. Depending on the latitude, this optimal temperature may occur at different times of the year. As you might expect, clam reproduction takes place much earlier in the year in the southern latitudes than it does in the northern latitudes.

By the same token, clam growth and development is equally dependent on ambient water temperature. For example, clams in Florida tend to grow three times faster than clams in more northern climes. Florida clams can progress from a little more than a half inch to nearly three inches in diameter in about two years. For northern clams, it would take about three years to reach that size.

The second critical condition that must be present for successful fertilization to take place is an abundance of food. You see, in the best of circumstances, the surrounding waters should be chock full of planktonic food. That's because young clams need quick and ready access to food—lots of it. A plankton-rich environment ensures the development of clam larvae at a most critical time in their life cycle.

Phytoplankton, or clam food.

Since a successful spawning is highly dependent upon both water temperature and the availability of food, it's important to remember that these are not always environmental constants. There are often peaks and valleys—times when clams are ready to release their sperm and eggs, and times when they are not. In clam language, these are known as "pulses"—the most opportune times to release gametes. Depending on the region and latitude, these spawning periods may occur several times during the year. In fact, southern populations of clams may experience multiple spawning opportunities—dimodal (two peaks) or polymodal (multiple peaks). In some places the cycle may begin in early April and continue on until late September. By releasing their gametes during these pulses, the clams are able to minimize the amount of wasted cells, that is, cells released when water temperature and food availability are insufficient for growth and development.

When all the conditions are right, then it's time to party (clamwise, that is). First, male clams release semen through their feeding siphons. The semen, which contains pheromones, is spread through the water by currents and is subsequently ingested by female clams. Those pheromones stimulate the females to release their eggs, which they expel through their excurrent (outward flowing) siphon. Each egg, supported by a gelatinous envelope, floats freely in the current until it is eventually joined by free-swimming spermatozoa. However, it's important to keep in mind that the union of sperm and egg is often by chance. Mother Nature is fickle, and just because clams produce a lot of eggs and a lot of sperm doesn't mean that all those gametes get together, reproductively speaking.

Most of the free-floating gametes will become food for all sorts of other creatures. For some sea creatures, the release of sperm and eggs into the water is cause for celebration; it's a feast of gigantic proportions. When you consider that, on average, a single mature female clam will release millions of eggs per spawn,* then you can only imagine what the surrounding waters must look like to a hungry fish swimming through this smorgasbord of gametes and fertilized eggs.

Within twelve to fourteen hours, the eggs that are fertilized (and uneaten) hatch into microscopic creatures called "trochophores." A single fertilized egg will typically float along on the current. In less than a day, this larval form transforms into a veliger, a free-swimming animal that resembles a butterfly. Veligers have tiny lobes or wings that propel them through the water, (Imagine a large flock of underwater butterflies.) The foot, shell, and body organs begin to form during this stage, which lasts about six to ten days. The veligers feed on all the planktonic algae that must also be present in the water during this critical time period.

Unfortunately, very few clams survive beyond the veliger stage. Veligers are often eaten by predators, killed by unfavorable water temperatures or salinities, or carried by currents to areas with insufficient food.

FAST FACT: Geoducks (who you first met in chapter 5 and will meet again in chapter 9) may be the reproductive champions of the animal kingdom. A female geoduck can release up to 10 million eggs in ten minutes. A male geoduck releases enough sperm to fertilize 5,000 females (that's 1.5 billion eggs).

For the very few veligers that do survive, they will eventually drop to the sea floor and shed their winglike lobes. As soon as they touch bottom, they will each send out thin filaments (byssal threads) to hold themselves in place in the sandy or rocky substrate. Then, as these tiny clams mature,

---

* For comparison purposes, a human female typically has about one million eggs, all formed before birth. By the time of puberty, only about 300,000 remain. Of these, only about 300-400 will be ovulated during a woman's reproductive years.

a muscular foot replaces the filaments. The clam uses that foot to wedge or wriggle itself down into the sediments. (With some species this may be up to three feet deep.) With most clam species, you can only see its two siphons protruding up into the seawater.

The reproductive life of a clam is unique. Yet, for the millions of eggs and larvae that cloud shallow tidal waters, only a miniscule number survive. The ones that do, however, may live for thirty years or more.

As I'm sure Professor Shadle would have agreed, clam reproduction is considerably quieter than copulating chipmunks or fornicating ferrets, but certainly no less dramatic.

Chapter 8

Growing Up, Growing Old

The trick is growing up without growing old.
— Casey Stengel, MLB outfielder and manager

I T WAS AUGUST OF 1954. THE TRYOUT DATES FOR THE forthcoming Broadway production of the musical *Peter Pan* were being held at the Philharmonic Auditorium in Los Angeles. The production was in LA for two months of rehearsals before moving on to Broadway for a limited run of 152 performances.

My grandmother thought it would be a proper entertainment for a young lad of seven to take in this literary tale—a story with its roots firmly anchored in England and its direction under the most capable guidance of Jerome Robbins. It was to be my introduction to the world of culture and art, something that certainly beat all those boorish programs appearing on the new living-room contraption known as a television set. (Although my grandmother was hopelessly addicted to the bubble-infused, champagne bottle popping, "wunnerful, wunnerful" *Lawrence Welk Show* on KTLA-Channel 5—a program she watched religiously every week.)

FAST FACT: During the 1958–1959 season, Lawrence Welk's show was the first TV program to air in stereophonic sound. (Interestingly, stereophonic TV would not become the standard for television broadcasts for another twenty-five years.)

And so, "Gann" (as I called her) and I made our way to the Philharmonic one Sunday afternoon, she dressed in her "Sunday best"—white gloves, high-collared dress, proper hat, and handbag; me in a very uncomfortable suit, an unnecessary tie, dutifully polished shoes, and freshly pomaded hair. I distinctly remembered entering the theatre and seeing posters featuring actress Mary Martin soaring over the stage or stopped in mid-play crooning one of the classic musical numbers ("I've Gotta Crow," "Never Never Land") from the show.

While folks today are more familiar with Disney's animated version (originally released in 1953), the stage production has been performed in England since 1904 and in the United States since 1905. The most successful version, and the one seen by most people, was the musical version I saw with "Gann," the one initiated in 1954.

FAST FACT: Although Peter Pan is a male character, he has always been played on stage by female actresses (Mary Martin, Sandy Duncan, Cathy Rigby).

What many people often forget is that the full title of J. M. Barrie's classic story is *Peter Pan; or, The Boy Who Wouldn't Grow Up*. The basic theme of the book was centered on Peter's refusal to make the transition from the innocence of childhood to the responsibilities of adulthood.* He continues to play, to fly, and to get into minor skirmishes with other characters in a place appropriately known as Neverland. Throughout the story he encourages others to retain their childhood passions and to eschew the responsibilities adults must contend with on a daily basis. But it is the first line in the book that ultimately sets the stage for the ultimate fate of the characters: "All children, except one, grow up."

That may hold true for humans . . . but for clams—not so much!

Basic fact of life: most children will grow up. They will have temper tantrums, learn to walk, suffer through bouts of acne, learn to drive (and parallel park), go to school, have heartbreaks, meet that "certain someone," get married, pay taxes, incur a hefty mortgage, raise their own kids, pay more taxes, retire and constantly complain how life was in "the good old days." According to the World Health Organization, average life

---

* One of my favorite quotes about this inevitable process was penned by playwright David Mamet: "Old age and treachery will always beat youth and exuberance."

expectancy (throughout the world) at birth in 1955 was just forty-eight years; in 1995 it was sixty-five years; in 2025 it will reach seventy-three years. In the United States (as of 2013) the average life expectancy was seventy-nine years overall (seventy-six years for men and eighty-one years for women).* In short, if you have kids, the chances of their reaching adulthood are pretty good.

For clams, the chances are slim to none!

Clam Development Stages

As you will recall from the previous chapter, when the time is right (or, when the moon is bright), all the boy clams release their mature gametes (sperm) into the water and all the girl clams release their mature gametes (eggs) into the same expanse of water. This is known as broadcast spawning. You may also remember from our discussion of giant clams, a transmitter substance called Spawning Induced Substance helps synchronize the release of sperm and eggs (at approximately the same time) to ensure fertilization. Clams can detect SIS immediately with their cerebral ganglia (similar, I suspect, to how a teenage boy can detect when his date has applied a liberal quantity of "Intense Flaming Desire and/or Lust" perfume.). As soon as SIS is sensed, a vast army of clams is sufficiently stimulated to fill their water chambers and vigorously contract. After several contractions, all those clams begin to release vast and enormous clouds of sperm and eggs into the water. This is known as the "ripe stage" in clam reproduction.

Think about this: A single healthy female clam is amazingly fecund. She may release anywhere from sixteen million to twenty-four million eggs in a single spawn, and, with repeated spawnings, individuals may release more than sixty million eggs over an entire season. That's literally buckets and buckets of eggs. Some of those millions and millions of eggs will be fertilized, but most will not.

The eggs, spermatozoa, and clam larvae all become part of the zooplankton that may ebb and flow through a shallow water environment—an environment, which in summer, can provide a nutrient-rich soup for a wide variety of animals. Many scientists note that a shallow seaside estuary (particularly during the summer months) may harbor as many as thirty

---

* Interestingly, the United States only ranks 26th in terms of overall life expectancy. The top five countries are Japan (83 years), Switzerland, San Marino, Italy, and Singapore. The bottom five countries (out of 190), in terms of overall life expectancy are Mozambique, Lesotho, Zambia, Angola, and Swaziland (31.88 years).

million clam larvae per square meter. However, as you might expect, mortality is very high during this stage of the clams' life. Most of the larvae are eaten by fish, crabs, birds, whelks, sea stars, and other mollusks. Even clams eat their own eggs and larvae by filtering them out of the seawater. Sadly, less than one in ten fertilized eggs survives.

You'll recall that the surviving fertilized eggs develop into veligers about twenty-four hours after fertilization. The larvae grow to a maximum size of 200 to 275 micrometers (approximately 0.00984 inches, or just slightly smaller than a single celery seed).

By the sixth to tenth day, the skin-like outside tissue, called the mantle, initiates a remarkable biological and chemical process, calcification, and the clam begins to build its permanent home, the shells. This is a universal process practiced by familiar marine organisms of various stripes: clams, oysters, coral, abalone, sea stars, sea urchins, and barnacles, to name a few. Less well-known organisms—such as coralline algae, brachiopods, coccolithophores, and several types of pteropods—are also calcifiers.

You see, seawater contains a wide variety of dissolved minerals;* chief among them is calcium carbonate. Calcium carbonate can be found throughout the terrestrial world in the form of limestone and other related materials (e.g., chalk). When those materials are eroded through the action of wind and water, particles of calcium carbonate are dissolved in streams and rivers that empty into the sea. Calcium carbonate, when dissolved, is ionized such that it is attracted by certain proteins.

FAST FACT: Much of the world's limestone deposits comprise the skeletal fragments of marine organisms, such as coral or foraminifera (of which there are about 275,000 species). Many of those deposits were laid down in ancient times when inland seas (such as the Western Interior Seaway that penetrated the heart of North America during the Cretaceous Period†) covered vast tracts of land.

Clams, just like other calcifiers, are able to create a layer of protein that attracts and attaches to the ionic calcium carbonate in seawater. Successive

---

* Chemical analyses have shown that seawater contains about 3.5 percent dissolved solids. More than sixty chemical elements have been identified, including salt, potassium, magnesium, phosphorites, gold, tin, titanium, and calcium carbonate.

† All of Colorado, for example, was under water about 100 million years ago.

layers of calcium carbonate are attached to the protein along the open edge of a forming shell. As more and more layers of calcium carbonate are attached to the protein, new layers of protein are created on top of each calcium carbonate deposit (layer), thus attracting more material. With the extra weight of the shell, larvae no longer swim freely and settle to the bottom.

When they are large enough (approximately 0.2 mm), the larvae change into the juvenile stage by attaching to the substrate with thin byssal threads (like a spider spinning a web, only simpler). Clams don't burrow at this stage; instead, they wrap their byssal threads around sand grains or shell fragments. The attachment is not permanent, however; a clam can release itself and move on to a different habitat more to its liking, should it prefer (like a college student changing apartments multiple times during the semester).

After twelve to fourteen days, the surviving larvae metamorphose into seed clams, and they finally begin to resemble little clams. They eventually develop a foot, and it is at this time that they often burrow into a suitable substrate where they remain mostly immobile. Just like four-year-old boys, most species of clams prefer a combination of mud and sand as a suitable substrate in which to live. Depending on the species, other suitable substrates may include pure sand, grave, and mud. For the most part, a clam's favorite habitat is a shallow estuary where the current moves the water at a fairly leisurely pace. Clams don't like fast-moving water (food spins by them too fast), nor do they like turbid waters. Since they feed exclusively on suspended food particles, excessive turbidity can clog their filtering system and eventually kill them.

There's another critical element important to the growth and development of clams: water of the proper salinity. Clams grow best in seawater that contains about 20–30 parts per thousand (ppt) of salt.* In fact, a larval clam will not begin its metamorphosis to mature clam unless the salinity is at least 18–20 ppt, ensuring that the seed clam will not settle in an area where the salinity is unsuitable for adults.

When a seed clam finds an environment with proper salinity and adequate food supply, it becomes a permanent resident. The clam opens its shell, extends its short, hatchet-shaped foot, and uses it to dig into the substrate. The clam burrows under approximately one inch, and then extends its pair of siphons upward to draw in seawater and to expel waste. At a shell

---

* Seawater salinity varies throughout the world. The salinity of the open ocean averages about 30-35 ppt. On the other hand, the Red Sea has an average salinity of 40 ppt, while the Baltic Sea has an average salinity of only 8 ppt.

length of about seven to nine millimeters, the byssal gland is lost and the young clam becomes a certified juvenile.

> **FAST FACT:** If you are interested in seeing a clam's foot (as well as learning how to dissect a clam in the comfort of your own kitchen), check out the following website: www.biologyjunction.com/clam_dissection.htm.

Adulthood

Once the clam has found its home, it will not go far from its territory. Although a clam can use its foot to dig itself out and to relocate, most marine biologists believe adult clams usually stay within one square yard of their home. However, if a clam senses danger (such as the footfalls of an author writing about said clam), it will burrow deeper into the substrate until the danger passes.

In warm, temperate areas, the clam's most significant growth will occur in spring and fall. That's when optimum water temperatures coincide with the availability of food. Growth tends to decrease through the summer months and virtually ceases in the winter (particularly at water temperatures less than 48°F). In those areas with longer growing seasons (Florida, for example), clams may reach market size in a shorter period of time than they would if they were planted in New England substrates. As a result, growth rates may range from two and a half years in some locations to as much as eight years in others before the resident clams reach harvest size.

However, just like with you and me, growth isn't constant; it tends to decrease with age. In short, the development of clams is not a linear progression. For there is one singular process that enters into the picture whenever we discuss our favorite bivalves. That is, sexual differentiation.

While the sex of humans is determined at conception, for clams it's not so simple. You see, as clams grow, they often undergo a sex change. At one year old, most clams are functional males (they can do all the things that male clams can and should do). Even though they are initially males, their permanent sexual identity is not yet determined. That's because, over the next year or two, as they mature, they can remain a male or become a female—a process that occurs at about a one-to-one ratio.

I'll not go into all the biological and physiological process at work here (to prevent this book from being tagged with an "R" rating). Suffice it to say, many species of clams have the capacity to change their sex. To biologists this is known as sequential hermaphroditism. You might find it interesting, as I did, that in the animal world, sequential hermaphroditism (when the individual is born one sex and changes sex at some point in their life) is not as uncommon as you might expect. Animals that do this include those who change from male to female (e.g., clownfish, some flatworms, and the black swallowtail butterfly) and those who change from female to male (e.g., blue-headed wrasses, California sheephead, parrotfish, and certain frogs).*

As adults, hard-shell clams will eventually grow to a maximum shell length of about four and a half inches. And, depending on the species and local living conditions, most hard clams will have a life span of approximately three to ten years. Soft-shell clams, on the other hand, reach a maximum shell length of four inches at maturity. They tend to live slightly longer than hard-shell clams, reaching ages of ten to twelve years.

> **FAST FACT:** The smallest clams in the world belong to the *Sphaeriidae* family. These clams live in freshwater lakes, streams, rivers, and ponds. Their shell length ranges from 0.039 to 0.98 inch, with an average length of approximately about 0.23 inch, or the size of a popcorn kernel. They are short-lived, however, with a life span of about one to three years.

Although life as an adult is somewhat less strenuous than life as a juvenile, that's not to say that that life is easy. Besides human predators, clams must survive a whole raft of potential consumers. These include whelks (whom we met in chapter 6), moon snails, and oyster drills—all of whom can easily penetrate the clam's armor to reach the tasty animal inside. Other enemies include blue and stone crabs (watch out for those powerful crushing claws), puffer fish, drums, sea trout, skates, and rays. For most adult clams, their best defense is their size. In essence, the larger the clamshell, the better the defense. Although clams can burrow deeper in the bottom when danger threatens, having a large, thick shell may be the best protection of all.

---

* Certainly one of the most unusual animals, sexually speaking, would be the male mourning cuttlefish. This creature has the unique ability to split itself down the middle, appearing to be male on one side, and female on the other.

Lifespan

For most people, their age is something they would rather hide from others, something one never discusses with casual acquaintances, or a fact of life that is seldom, if ever, revealed. Most folks would rather dismiss or "forget" their actual age—a psychological reaction, I suspect, tantamount to admitting you were senile, infirm, and incapable of making rational decisions. Much of that reaction may be due to today's "youth culture" in which a youthful appearance is celebrated in the products we buy ("Instantly relaxes the look of wrinkles and improves the appearance of deep eye lines in 14 days!"), the "news" we read in supermarket tabloids ("Snag a sexy guy? Here's how to do it!"), or the celebrities we worship ("[name of favorite hunk] treats us to another smoldering, shirtless photo!").

Obviously, we live in a society dedicated to fibbing—especially when it comes to our age. This is not something new in the human experience; it is a cultural constant that has been practiced over the millennia and in every civilization since the dawn of recorded history. Witness the search for the "Nectar of Immortality" as described in ancient Hindu scriptures, sixteenth-century Spanish explorer Juan Ponce de León and his never-ending quest for the proverbial Fountain of Youth (which is in St. Augustine, Florida, by the way), and Oscar Wilde's classic novel *The Picture of Dorian Gray*, about a man willing to sell his soul to remain constantly youthful, among numerous examples in our eternal quest for eternal youth. Despite the efforts of AARP, we are awash in stories, legends, and tales about our never-ending need and constant longing for everlasting youth, rather than a satisfying acceptance of age . . . specifically, old age.

For many folks—particularly those of us well into our fourth or fifth or (ahem!) sixth decade—talk of time, or of age, is frequently difficult. Although humans may be culturally predisposed to lie and fib about their age, animals cannot. For example, consider the American sand-burrowing mayfly (*Dolania americana*). This creature has what we might term a very short and very hectic life. Its entire reproductive life takes place in less time than it does for you to read the front page of your daily newspaper or consume a Big Mac at your local fast food restaurant. After its final molt, and in just about five minutes, the female of this species must locate a suitable mate, copulate, and then return to its watery origins to lay a clutch of eggs. Then it dies. Its entire adult life is dedicated to one and only one function: preservation of the species.

FAST FACT: Other short-lived animals include the pine procession-ary moth (*Thaumetopoea pityocampa*), which lives for twenty-four hours; a wasp (*Trichogrammatoidea bactrae*) that lives about twenty-eight hours; and male gall midges of the genus *Rhopalomyia*, which emerge as adults in the morning and are dead by noon.

Yet, if we were to take a trip toward the other end of the longevity scale, we might discover bowhead whales. These sixty-foot, sixty-ton levia-thans inhabit Arctic waters year round and have heads that are up to 40 percent of their body length. Through an analysis of amino acids found in its eye lenses, scientists determined that one whale, killed by Alaskan natives in 2000, was 211 years old at the time of its death. In other words, it was swimming around Alaska when Thomas Jefferson was president (1801–1809).

And, then, there's the rougheye rockfish, a creature that proves you don't have to be pretty to live a long life. Indeed, with up to ten spines along the lower rims of its eyes, an oversized mouth, and a bright-orange color, the rougheye rockfish looks like a badly designed alien from a 1950s sci-fi movie. Yet one specimen recently checked in at an impressive 205 years of age.

FAST FACT: Research conducted at the University of California, Irvine, suggests that certain species of the hydra (a marine creature found in oceans throughout the world) is "capable of escaping aging by constantly renewing the tissues of its body."

Lifespans are, at best, averages. They are estimations and approxima-tions of how long a typical individual or entire species will live. Yet, it is the extremes of a species' lifespan that fascinate us most. Extremes defy averages, extremes challenge the status quo, and extremes give us all something to shoot for. If humans are expected to live until 85 and one of us lives to the ripe old age of 110—that's exciting! If domesticated cats have an expected life of 12 years and your pet cat Fluffy lives to be 24—that's exciting! If the average life expectancy of a hippopotamus is 30 years and one lives to be 55—that's exciting (but probably only to another hippo)!

Now, imagine being around when Shakespeare was writing his comedies. Consider living during the same time the English were establishing their first colonies in the Americas. Think about growing up when Galileo was being tried for advocating the heliocentric model of the universe. Imagine being around when explorers such as Francisco Pizarro, Hernando Cortés, and Christopher Columbus were crossing the oceans and continents of new worlds to search for riches and lands to conquer. Imagine living since before the invention of electricity, since before industrial factories, and since before antiseptics and antibiotics.

Shakespearian comedies . . . colonizing the Americas . . . Galilean physics—well, there's a certain Icelandic clam that lived through all that . . . and more!

Also known as the Icelandic cyprine, ocean quahog, and mahogany clam, *Arctica islandica* is a long-lived, suspension-feeding, bivalve mollusk. It lives burrowed in the top two inches of sand and muddy substrates of near-freezing habitats around the shelf seas of Northern European and North American continents. Its range extends from the Bay of Biscay off the west coast of France to the subarctic waters around Iceland, where it inhabits depths between 80 and 260 feet below sea level. The species is dioecious (having male reproductive organs in one individual and female in another), with larval development taking between thirty and sixty days depending on seawater temperature. Sexual maturity is reached between seven and thirteen years of age.

The anatomy, behavior, physiology, and, more recently, the ecology of *Arctica islandica* has been extensively studied because of their commercial importance along the American east coast: 165,000 tons are collected globally each year, principally by hydraulic clam dredges. It was only recently, however, that researchers examined the scientific literature. They found references to specimens with ages between one hundred and two hundred years old, as well as information on clams collected in the North Sea and near Iceland attaining ages far beyond the seemingly old-age limit of two hundred years.

Then, in the 1980s, scientists made a fortuitous discovery: they found an extremely long-lived *Arctica islandica* specimen (a 374-year-old clam housed in a German museum) that was far older than was previously thought possible. This discovery was the stimulus the scientists needed to begin investigating other clams—specifically how the clams "recorded" their years of existence. In a stroke of scientific luck, it was discovered that, just like many land-based trees, clams recorded their lives with rings. These rings were deposited like clockwork—that is, one ring for every year of existence.

"Ah-hah!" the scientists said.

Unfortunately, they couldn't do anything with that "Ah-hah!" (at least, not at that time). Then, in 2006, a team of scientists from Bangor University in England was conducting research as part of a long-term project to understand how the world's climate has changed over the past several centuries.* One day, while anchored off the frigid northern coast of Iceland, they pulled up what may be the oldest animal on record—a relatively undistinguished clam. This particular *Arctica islandica* was among a haul of three thousand empty shells and thirty-four live mollusks eventually taken to the laboratory.

The clam was turned over to sclerochronologists, the folks who study the growth and age of clams using those annual growth lines in the shell. These scientists do their counting in much the same way as dendrochronologists study tree rings. Thus, you can only imagine the surprise and amazement ("Ah-hah! AH-HAH!") when the investigators methodically counted the clam's growth rings and came up with a figure of 405 years. The clam was nicknamed "Ming" after the Chinese dynasty that ruled when the clam "settled"—the mollusk equivalent of being born. The scientists were dancing in the street.† (As you will note, some scientific discoveries are simply a matter of chance and luck, as well as being in the right place at the right time.)

Unfortunately, the clam died during the ring-counting process.

Since its extraordinary discovery, this exceptional clam has been tagged with the designation "tree of the sea." That's because the growth-increment series measured from its shells could be used (retrospectively) to reconstruct marine environmental change. You see, the shells of *Arctica islandica* contain an ontogenetic (the origin and development of an organism) record of shell growth in the form of wide annual summer growth increments separated by narrow growth lines. Thus, if you were to count the number of lines in this clamshell (or any clamshell), you would get an accurate estimate of the age of the clam. You would also be able to determine, through a precise measurement of increment width, the clam's inter-annual growth rate.

---

* The Bangor University School of Ocean Sciences is a world leader in biological chronology and has been working toward constructing a thousand-year time scale for the marine environment using the growth increment series in *Arctica islandica* shells.

† In its 2007 yearbook, *Time* magazine designated this discovery as one of the "Top Ten Scientific Discoveries" of 2007. (It actually came in as #9.)

> **FAST FACT:** Each year, a shell layer as thin as 0.00394 inch is laid down by the *Arctica islandica* clam.

The researchers hope to use this and other shell studies to reconstruct a record of environmental changes over the past several centuries. Al Wanamaker, assistant professor of geological and atmospheric sciences at Iowa State University and a team member at Bangor University, said he believed that the aforementioned clam had survived so long because fisheries and predators were so few in the region where the clam lived. In some areas, clam populations have been wiped out through overfishing, while marine predators, including cod, seals, and wolf fish, also take a hefty toll. Wanamaker added, "Its death is an unfortunate aspect of this work, but we hope to derive lots of information from it. For our work it's a bonus, but it wasn't good for this particular animal."

Chris Richardson, also a professor of marine biology from Bangor University's School of Ocean Sciences, told the BBC shortly after determining the clam's age: "The growth-increments themselves provide a record of how the animal has varied in its growth-rate from year to year, and that varies according to climate, sea-water temperature and food supply."[*]

"And so by looking at these mollusks we can reconstruct the environment these animals grew in. They are like tiny tape-recorders, in effect, sitting on the seabed and integrating signals about water temperature and food over time." Richardson said the clam's discovery could help shed light on how some animals can live to extraordinary ages.

"What's intriguing the Bangor group is how these animals have actually managed, in effect, to escape senescence,"[†] he said. I find it most interesting (and slightly ironic) to note that the university received £40,000 ($67,472 in 2014) from the United Kingdom charity "Help the Aged" as part of its funding for this research project.

However, the question still remains: "Why do these clams live so long?" The Bangor scientists are attempting to find out and believe the clams may have evolved exceptionally effective defenses that hold back the destructive

---

[*] "Ming the Clam Is Oldest Mollusc," *BBC News*, October 28, 2007.

[†] The term "negligible senescence" was coined by Caleb Finch, professor of gerontology and biological science at the University of Southern California, to describe very slow or negligible aging. He listed several animals with this characteristic, including vertebrates (sturgeons, bowhead whales, turtles, rockfish) and invertebrates (hydra, sponges, coral) and specifically named *Arctica islandica* as a potential organism.

aging processes that normally occur. "If, in *Arctica islandica*, evolution has created a model of successful resistance to the damage of aging, it is possible that an investigation of the tissues of these real-life Methuselahs might help us to understand the processes of aging," explains Richardson.

"It's quite possible others are out there in the water that are six hundred years old," proffered Wanamaker.

Then, in 2013, something amazing happened - all the "Ah-hahs!" turned into "Oh, damns!" You see, the scientists who first counted the rings on "Ming" stated that they actually got it wrong the first time around (a resounding chorus of "Oops"). During the original count, the researchers counted the growth rings inside Ming's hinge ligaments. This was done because the rings there are better protected; less subject to the wear and tear that might occur on the outside of the shell.

However, that ring-counting process was exceedingly difficult simply because all the rings were packed so close together. So the scientists decided to do a recount—carefully calculating the number of rings on the outside of the shell. To their amazement, they discovered that Ming was actually 507 years old, not 405 as originally thought. "We got it wrong the first time and maybe we were a bit [hasty] publishing our findings back then. But we are absolutely certain that we've got the right age now," said scientist Paul Butler.

Think about this: At 507 years old, Ming lived from 1499 until the day scientists opened up its shells (in 2006) to count its rings. It was "born" in the same year that Switzerland became an independent state (September 22, 1499), just before Pedro Álvares Cabral officially discovered Brazil and claimed the land for Portugal (April 22, 1500), and just after Columbus's third voyage to the New World (May–August 1498).

Wow!

Just imagine having lived for five hundred years!

Just imagine having survived for fifty decades!

Just imagine having endured five centuries of brutally frigid and violently tempestuous North Atlantic waters with absolutely no opportunity to travel to sky-blue tropical seas placidly shimmering under invitingly warm Caribbean sunshine!

Just imagine!

PART III

Business and Pleasure

Chapter 9

It Came from Beneath
the Sea

A lot of people attack the sea; I make love to it.

—Jacques Cousteau

B Y NOW, YOU ARE WELL AWARE OF MY YOUTHFUL predilection for B-grade science fiction movies. I was particularly enamored with movies featuring stop-motion animated monsters from far-flung galaxies, prehistoric worlds, or even from the deep waters off the California coast. These were classic cinematic marvels created long before special effects became the province of computer nerds and multi-million-dollar movie budgets.

One of the most popular movies of the fifties (and my childhood) was *It Came from Beneath the Sea* (1955), which was distributed on a double bill with *Creature with the Atom Brain*. (As an eight-year-old, I was in science fiction heaven.) *It Came from Beneath the Sea* was designed to showcase the special model-animated effects of Ray Harryhausen. It was made on a total budget of $150,000 and had a running time of seventy-nine minutes, both paltry, nay miniscule, by today's cinematic standards. The poster for the film showed a gargantuan octopus ("Out of primordial depths to destroy the world!"), its writhing arms savagely wrapped around the Golden Gate Bridge and the normally laid-back citizens of San Francisco screaming and dashing off in all directions to escape this atomic-fortified beast bent on destroying the sixth most visited tourist destination in the United States.

It Came from Beneath the Sea stars a hellish creature that rises up from the depths of the Pacific Ocean when radiation from H-bomb testing affects its normal feeding habits. Not only does it develop an insatiable hunger for all manner of maritime vessels, it becomes highly radioactive, thus making it virtually indestructible by the current armament of the US Navy (who often appear so inept in these films). The lure of San Francisco and the prospects of wrapping its arms around the Golden Gate Bridge propel this atomically fueled octopus towards an inevitable clash with military and civic authorities throughout the Bay Area.

An atomic submarine captain (played by Kenneth Tobey) teams with a pair of scientists (Faith Domergue, Donald Curtis) in an attempt to prevent the monster from terrorizing "The City by the Bay" as well as other potential "octopus-prone" cities along the West Coast. Nevertheless, the oversized cephalopod, in its Pacific Ocean rampage, snacks on several freighters and conveniently dines on a few unlucky tourists who happen to be on the wrong Oregon beach at the wrong Oregon time. After a giant suction imprint is found on the beach and the local sheriff is unceremoniously whisked away by the creature (obviously, no respect for the law), it becomes apparent to any remaining skeptics that this is no ordinary invertebrate.

> **FAST FACT:** Because of severe budget limitations, Ray Harryhausen was required to build his creature with only six tentacles, instead of the usual eight. He laughingly named it "The Sixtopus."

In quick succession the Navy mines the entire Pacific coast, an electrified safety net is strung underwater across the entrance to San Francisco Bay, and the Golden Gate Bridge is wired for a high-voltage greeting. It is apparent that the creature is in for "the shock of its life." As the ultimate *coup de grâce*, one of the scientists reveals a special jet-propelled atomic torpedo—a weapon that will be shot into the creature, driving it out to sea before the torpedo's detonation (and the fatal obliteration of the constantly writhing monster).*

---

* As kids we didn't even consider the unbelievable irony of just happening to have an atomic weapon standing by in case a humongous (and quite enraged) creature decides to visit your town, consume your citizens, or destroy one of your beloved structural icons.

The Navy quickly orders the Golden Gate Bridge abandoned—a good move since the sight of orange-painted metal structures spanning large bodies of water apparently enrages the creature to no end. Yet, in spite of all the precautions, the creature methodically proceeds to dismantle the bridge, the residents of San Francisco panic and beat a hasty retreat down the peninsula, and authorities struggle to evacuate several nearby buildings. In quick succession, a helicopter is wacked from the sky, the Ferry Building is lovingly embraced and then demolished by the creature, and an enormous tentacle brings havoc and destruction down the length of Market Street (sending hyperventilating attorneys into a frenzy of litigation).

It is apparent that this is a creature with attitude!

As the movie's theme music intensifies, so does the action. Tiny soldiers with flame throwers encourage the enraged octopus (or at least its tentacle) to return to the sea, the creature grabs a submarine, a scuba diver shoots an explosive charge into the monster, and the octopus is harpooned in the eye by another diver. Apparently tired of all these humanoid shenanigans, the octopus releases the submarine and heads for the open sea. It is then (cue the dramatic, high-tension music) that the torpedo is detonated, resulting in the complete and total annihilation of the giant octopus (much to the chagrin of teuthologists* everywhere). In the end, the three main characters (apparently disturbed, but not harmed, by the creature) retreat to a nearby cocktail lounge to toast their oceanographic success. And, as we expected, soft violins play and they all live happily ever after.

> **FAST FACT:** Many biologists consider octopuses to be highly intelligent creatures. In fact, octopuses can navigate their way through mazes, engage in sophisticated problem-solving, use both long-term and short-term memory, and handle various types of tools.

While my youth was filled with any number of movie monsters, critters, and beasts from the ocean, I was also aware of the riches of the Pacific. Since I lived in an oceanfront environment and regularly dined on seafood of every epicurean design and description, I (and my family) was inexorably tied to the sea in so many ways. Thus, both my cinematic and gastronomic pleasures were frequently satisfied by one sea inhabitant or another—both the large and the small.

---

* Teuthologists are scientists who study cephalopods (squids, octopuses, cuttlefish, and the nautilus).

Harvesting Clams

As you know, many creatures (both good and bad) come "from beneath the sea." For some of those critters, the challenge is getting them out of the sea and into pots, bowls, and large mugs of melted butter. As a result, clam harvesting has been a passion of humans since long before recorded history. In the old days clams were pulled from beaches by hand, a labor-intensive activity that required several people to scour beachheads for long periods of time in order to obtain a sufficient quantity of clams for a forthcoming family dinner or a village's religious celebration. In so many ways, this was backbreaking and required a lot of time and patience. Yet, clam harvesting is still practiced today, although now as a satisfying recreational activity—one we will visit in considerably more detail in chapter 11.

While clamming at low tide was the only way to obtain clams by primitive people, it became quickly apparent there were even more clamming possibilities in the shallow waters of bays and harbors—especially along the sandy bottoms that could not be reached by hand. What eventually evolved was clamming from specialized boats, often with flat bottoms to negotiate shallow waters. Clammers would go out on these boats with simply constructed, though quite efficient, tools, such as clam rakes.

As an example, imagine a garden rake with a telescoping handle, extra-long tines, and a basket-like cage attached behind the tines. The clammer stands in his or her boat and drags the clam rake over the sandy bottom directly around the boat. Clams are scratched up from the substrate and into the small wire basket. Using this particular clam rake, the clams are pulled up, deposited on the deck, and the action is repeated. When one area has been raked clean, the boat is moved to a new area and the clammer goes through all the motions once again. (I don't know about you, but my shoulders ache just thinking about this.)

FAST FACT: There are probably as many different kinds of commercial clam rakes as there are professional clammers. One company (the R. A. Ribb Company of Harwich, Massachusetts) offers the following professional clam rakes: Rhode Island, Long Island "Bubble", Virginia Harvester, Suitcase, pocketbook, wireback, soft-shell, overboard, and hardshell. In addition, they also have eleven different styles of clamming rakes for us amateurs.

Another common tool used for bay clamming is specialized tongs. A clamming tong is similar to two clam rakes attached together, each with teeth hinged like scissors. A clammer will use the tongs to probe the sand for clams, then open and close the two handles and scoop up any clams unfortunate enough to be in the area.

Again, these methods of clamming, used by generations of clammers, especially along the eastern seaboard, require lots of labor and lots of time. For most clammers, profit margins are low, savings accounts unheard of, and expenses (salaries, boat, fuel) keep escalating. By now, you might be thinking that there must be a way to maximize clamming so that it would be more efficient, generate increased and continuous harvests, and result in lower mortality for the clams harvested. And, indeed there is!

It is the hydraulic escalator dredge.

The hydraulic dredge revolutionized commercial clamming simply because it made possible the harvesting of large numbers of clams from the sands in which they were buried. Instead of bushels of clams, clam harvesting could be measured in considerably larger units, such as tons . . . along with larger units of monetary profits, such as thousands rather than hundreds.

Fletcher Hanks, head of the Hanks Seafood Company in Easton, Maryland, developed one of the first successful escalator harvester designs* in the 1950s for use in the Chesapeake Bay. His apparatus was first used commercially in 1952 and has been adopted in numerous states, including Florida, Maine, New York, North Carolina, South Carolina, Oregon, Virginia, and Washington. It can be used to harvest soft-shell and hard-shell clams, as well as other shellfish.

An escalator dredge is a fifty-foot-long device that looks like the ladder on a hook-and-ladder fire truck. The dredge, or digger, is towed beside a boat (average length: thirty to fifty feet), along with a steel-mesh conveyor belt. When the boat arrives at an appropriate clamming area, the "ladder" is dropped into the water to run a scoop along the bottom. An oversize pump on the boat pumps seawater through a large hose to a manifold on the front of the dredge. Six to twelve feet below the water surface, jet streams of water blast into the sandy bottom at depths up to eighteen inches. This

---

* The patent application denotes it as an "Apparatus for dredging from the stern of marine vessels (US 4028821 A)—A dredge for operation from the stern of marine vessels and of the type having runners which enable it to be moved along a ramp at the stern of a marine vessel is provided with structure for preventing its twisting and misalignment. . . ."

action creates trenches or furrows in the sandy bottom and temporarily fluidizes the sand, which allows the dredge to pass through.

Spray from the hose kicks loose the sand and knocks unburied clams and other substrate materials (shells, debris, stones) onto the scoop. Because of the carefully set spacing of the bars making up the body of the dredge, most of the smaller clams and detritus commonly on the ocean floor pass through. The larger clams, however, do not.

From the scoop, the clams (and any other large items) are funneled onto a conveyor belt that carries them up to the boat. Two or three crew members at the other end, on the boat, rapidly separate the clams from any bottom material. Dead shells and other materials remain on the conveyors after culling and are returned immediately to the water.

Operation of harvesters usually proceeds without a set pattern, with dredge "tracks" crossing or recrossing each other many times. As the dredge is towed along the seafloor, it shoots jets of water into the sandy substrate, which makes the capture of clams in the dredge more efficient. One clam boat can dredge 1,200 to 1,300 square feet per hour and unearth thirty to one hundred bushels of clams per day.

Suffice it to say, mechanical dredges are not without controversy. Although there are some mechanical restrictions that the gear imposes— such as hose length and pumping pressure—the evolution of the basic design of these harvesters has produced machines that can efficiently and rapidly remove large numbers of bivalves from a given area. As a result, there is, as you might imagine, some considerable disturbance to seafloor sediments. A dredge's scoop and pressurized water sprayed from its hose can turn up a sandy bottom in no time and rake out clams (as well as other fauna and flora) by the hundreds. The dredge leaves behind a trench in the seafloor and suspends sediments.

Scientists have calculated that the sediment disturbance caused by the dredge takes approximately forty or so days to recover. Since hydraulic clam dredges cause high levels of disturbance, several environmental organizations (e.g., Seafood Watch) deem the impacts of hydraulic clam dredging on substrate to be high. Besides the loud diesel engines, the pungent odors, and the chance for leaking fossil fuels, it seems safe to say that the disruption to a shallow water ecosystem can be biologically devastating and frequently permanent.

Many states have enacted legislation to restrict and curb not only the actions of hydraulic dredges, but also the number of dredges that may operate within state boundaries. Some states also prohibit their use where they affect navigation in channels. For example, North Carolina has a per-bag limit on clam dredging to prevent excessive stripping of natural resources. Connecticut, on the other hand, requires no prior approval for use of dredges on leased or private grounds. In Alaska a state permit is required. In New York mechanical harvesting is permitted on leased grounds or for transplantation elsewhere. And in Maine a special license is required because of low clam stocks. As a result of these regulations, only a handful of commercial dredging companies are currently in operation.

> **FAST FACT:** Throughout the United States there are eight fishery management councils responsible for the management of marine fisheries. They are located in the following regions: New England, Mid-Atlantic, South Atlantic, Gulf of Mexico, Caribbean, Pacific, Western Pacific, and North Pacific. The Mid-Atlantic Council (which includes the states of New York, New Jersey, Pennsylvania, Delaware, Maryland, Virginia, and North Carolina) is the only council that sets commercial limits for clams (of any species).

Along the eastern seaboard, for example, the Hydraulic Clam Dredge Fishery conducts regular stock assessments of surf clams and quahogs. Unfortunately, not every state is able to conduct systematic and detailed assessments of their offshore clams. Nevertheless, harvests are often monitored by other agencies and environmental groups to help ensure stabilized populations.

A Family-Owned Shellfish Company

A dark bank of clouds scuds across the horizon, the parting remnants of last night's rainstorm. Sunlight forces itself around the edges, accenting the corridor of trees and the rich emerald life on both sides of the highway. I have left Seattle mid-morning and am now driving south on Interstate 5 through Tacoma and past Olympia, then north on Highway 101 as it ribbons its way up the Olympic Peninsula. My destination is the Taylor Shellfish Company, just below the tiny town of Sheldon. Taylor Shellfish is the biggest producer of shellfish in North America and I am scheduled for a clam farm tour with Marco Pinchot, the community relations and sustainability manager for the company.

> **FAST FACT:** Taylor Shellfish deals directly with numerous restaurants throughout the United States. In one week they will sell about 78,000 pounds of shellfish (up to 100,000 pounds during holiday weeks) for a total of up to 4 million pounds of shellfish each year.

Shortly after arriving at the company's headquarters, a sprawling complex set amidst acres of vibrant greenery, Marco and I exchange quick introductions and he departs briefly to finish off some essential paperwork due later that day. I am left in the capable hands of the enthusiastic and energetic Karen Underwood, a vibrant woman in charge of specialty sales for the company. With all the skill of an accomplished tour guide, Karen expertly leads me through a series of cavernous rooms, each chillier than the previous, and each humming with banks of enthusiastic workers shucking oysters, cleaning clams, packing mussels, grading geoducks, and filling orders ready to be sent over to Seattle or around the world. In each room, the atmosphere is one of occupational satisfaction—music fills the air, workers are engaged in spirited conversation, and orders are processed with both speed and alacrity. The camaraderie is palpable and the atmosphere is intensely positive as polyglot crews (I hear at least four different languages) attend to the myriad responsibilities associated with a vibrant and highly productive business.

As we make our way between rooms and buildings, Karen regales me with numbers, statistics, and a passion for the industry. She, like so many of the employees, is sincerely dedicated to this operation. She tells me that most of the employees are long term, many with twenty, thirty, or more years dedicated to this family-owned and family-run operation that places a high value not only on the products it sells but also on the employees who work there. It is quite evident that there is both passion and satisfaction here—a spirit that goes far beyond paychecks and the typical nine-to-five workday.

After an hour of chilly rooms (including one at −10°F), the click-clack of machinery, and the singular aroma of fresh seafood, Karen delivers me back to Marco's office. He and I hop into a well-used company van for a short trip down the road to one of Taylor Shellfish's Manila clam farms.

A shock of red hair, a neatly trimmed goatee, a pair of glasses, and the sincere enthusiasm of a dedicated conservationist, Marco Pinchot has been working for Taylor Shellfish for about six years. His last name may sound familiar to you—Marco is famed conservationist Gifford Pinchot's great grandson.

> **FAST FACT:** Gifford Pinchot served as the first chief of the United States Forest Service from 1905 to 1910. He was a strong conservation advocate and worked tirelessly at reforming the management and development of forests throughout the country. His legacy was the controlled, profitable use of forests and other natural resources. Pinchot was also the twenty-eighth governor of Pennsylvania.*

In his late thirties, Marco is a true evangelist for the outdoors—one who is dedicated, passionate, and sincere about the positive interactions that can exist between humans and the natural world. His family history portends an individual who works for environmental causes in a dynamic way—guiding folks to a better understanding of environmental sustainability and its role in their lives.

Marco grew up in upstate New York and in Connecticut. After attending high school in southern California, he eventually obtained a degree in marine biology from Evergreen State College in Olympia, Washington (Its mascot is Speedy the geoduck.) Later, he earned a master's degree in environmental education from Western Washington College.

For a while, he worked at the Lady Bird Johnson Wildflower Center in Austin, Texas. Then, a move back to the Pacific Northwest and the earning of a MBA brought him into contact with Bill Taylor, the owner of Taylor Shellfish Company.

Since then, he has worked for the company in public and community relations on issues as diverse as marketing, sustainability, and social responsibility. "I draw upon all my education for my job—communication skills, biology, and marketing," he tells me. It is clear that his words are both genuine and fervent. "This is a job that is constantly changing, constantly evolving," he says.

When I ask him about the greatest misunderstanding the general public has about aquaculture, he smiles and says, "Most people have this idea that wild is good and aquaculture is bad. That's, quite obviously, an

---

* My own connection to Pinchot lies in the fact that the 2,200-acre Gifford Pinchot State Park in Lewisberry, Pennsylvania, lies less than six miles from where I live. My wife and I often walk the trails and enjoy picnic dinners by the lake on warm summer evenings.

oversimplification. People tend to lump all forms of aquaculture together." As we discuss the issue, it becomes apparent that the general public often hears mostly about the "evil" aspects of the industry—public-radio stories about the easy spread of diseases among broods of penned tuna or newspaper accounts of farmed salmon in overpopulated tanks with little room to swim or breathe. As a result, most folks often get a distorted view of marine farming.

Marco reminds me of a quote from famed oceanographer Jacques Cousteau: "We must plant the sea and herd its animals using the sea as farmers instead of hunters. That is what civilization is all about—farming replacing hunting." We both agree that there are those who, in the name of aquaculture, will push the "environmental envelope" past its natural limits to make a fast buck. But, as Marco reminds me, "Fisheries may be the last place humans are hunting for animals." As we drive through a strikingly pristine environment, the two of us agree the statement is most powerful, particularly when everyone takes on the responsibilities of maintaining those fisheries, not for short-term economic gains, but for long-term sustainability.*

In short order we arrive at a small parking area astride Little Skookum Inlet (a branch of southern Puget Sound) and one of the many clam farms run by Taylor Shellfish. After strapping on my boots and making my way down to the shore, I am introduced to Claire LeVeque, the manager for the work crew that day. (Many of the crew wouldn't arrive until later.) Although Claire grew up in New York, she headed west to attend Evergreen State College and fell in love with the land. After college, she stayed on and has brought her passion for the outdoors to a job she absolutely loves— clamming. "Just last week, I saw eagles, ospreys, mergansers, and loons. Where else could I have a job with those kinds of encounters?" she exclaims with absolute conviction. For Claire, the best part of her work is the array of interesting people she meets and how the environment keeps changing over time. She does point out, however, that "[she is] in a male-dominated field, so that takes some getting used to."

A short time later, we are joined by Claire's boss, Brittany, and the two of them grab their hand rakes and buckets, walk over to a prime clamming area, and begin their work. As I was to learn, Little Skookum Inlet,

---

* According to an article on the National Geographic website ("Sustainable Food"), "Fish farms produce half of all the seafood the world eats—but not all of them are created equal. True sustainable operations minimize environmental impacts like pollution, disease, and other damage to coastal ecosystems on which wild species depend."

with its rich blend of mud and gravel, is an ideal area for the growth and development of Manila clams (whom we met earlier in chapter 5). The mix of gravel (the company occasionally refurbishes the area with loads of it) enables oxygen to get into the ground and eventually to the clams (which would make them, aerobically speaking, "happy as a clam").

Claire and Brittany dig for Manila clams.

> **FAST FACT:** Taylor Shellfish Company is the largest producer of Manila clams in the United States.

As I watch, it becomes abundantly clear that digging Manila clams out of the substrate is not a vocational choice for visiting authors with bad backs, an artificial knee, and a propensity to spend much of the day reposing in a cushy office chair in front of a computer. It's definitely not for those opposed to working on their hands and knees for long periods of time. But, Claire and Brittany do it with grace, enthusiasm, and an absolute sense of mission. Working together, they quickly fill a ten-gallon bucket with a clunky collection of medium-sized (and very marketable) clams.

Claire tells me that she usually works with a four-person crew and that it's not unusual for her crew to dig approximately 1,600 pounds of clams from the inlet every two days. Brittany confirms that number when she tells me that each person in the crew will extract about 200–300 pounds of clams per day on average.

It's quickly approaching low tide for the day, so Marco and I jump back into the van and set off down the road for a half-hour drive to a geoduck farm. After parking the vehicle in a copse of evergreens, we climb down a long hill and cross over a large expanse of muddy tidal flats. Here we find a crew of about a dozen men in the process of planting several thousand geoduck seeds in the oozy-mucky-soggy-muddy-slushy-boggy substrate.

Planting Geoducks

Geoduck farming is an unusual, distinctive, and curious venture, to say the least. Imagine, if you will, a vast muddy area (at low tide) of approximately four or five football fields in size. Arrayed across that enormous area are endless rows of PVC pipes sticking four to five inches out of the mud. (It looks like a plethora of tiny factory smokestacks—the aftermath of an enormous and muddy flood that has inundated the factories below.) And I mean rows upon rows upon rows. . . . It's as though an international plumber's convention were in town and all the plumbers stuck their excess PVC materials in the ground as some sort of special memorial or practical joke.

As we muck our way across the field and around the tubes, Marco describes the scene before me. He makes clear that geoduck farming is quite unlike the farming of any other animals or plants. It starts with a collection (hundreds of thousands) of those PVC tubes, each about nine inches long. The tubes are pushed about halfway into the substrate and arranged in long rows about one tube every foot. The rows are linear and square so that an orderly geometric pattern is preserved. From overhead it would look like a gigantic array of miniature missile silos poised to launch equally miniature projectiles into the stratosphere.

Geoduck nursery tubes.

Each of the workers is provided with a bucket of geoduck seeds—tiny critters about an inch to an inch-and-a-half long. As I hold several of them in my hand, it is as though I am looking at a collection of oversized peanuts, each with a calcium overcoat. However, these "peanuts" cost about fifty cents each and will eventually grow into delicacies that will sell (wholesale) for about thirty to thirty-five dollars each.

Using a long stick, each worker pushes three separate holes into the mud inside each PVC tube. One geoduck seed, siphon up, is placed about two inches down inside each of the three holes. Marco tells me, "The hope is that two of the three seeds will survive all the way to maturity so that we will end up with about two geoducks per square foot of the beach."

Geoduck seeds ready for "planting."

Then, after planting an expanse of tubes (now known as nursery tubes), the workers cover a large section of them with long, wide nets with half-inch to one-inch holes. The nets are then staked down on all corners to prevent them from being dislodged by the tides or by any large (land or sea) critters that may want to sample the wares. Marco informs me that at this point the tubes serve a variety of purposes. "One of the main purposes is that they create a little tide pool, which helps to keep the geoduck seed wet. [This is particularly important at low tide.] The clams don't like to go dry, so if you keep a little tide pool around them, they have a much better survival rate. Also the tube provides the nursery stock with refuge from predation. The chief predators include crabs, starfish, moon snails, diving ducks, and flounder—basically anything that would want to eat a geoduck."

FAST FACT: The seemingly innocuous moon snail is anything but. It is one of the most vicious predators in any marine environment. A moon snail will envelop its prey (a geoduck, for example) and begin to drill a hole through the shell using its radula and an acid secretion. Once through the shell, the snail will use its proboscis to consume the innards. If you've ever seen clamshells with a countersunk hole in them, blame the moon snail.

Marco tells me that for the next year to a year and a half, the baby geoducks will live inside the tubes, consuming the natural phytoplankton in the water as it ebbs and flows over them with every tide. Then, the netting and the tubes are removed and the geoducks repose (and continue to grow) in subterranean comfort for the next four to five years. At this point, it would be quite difficult to tell you were in the middle of an enormous geoduck farm, save for a random assortment of siphons poking up out of the mud squirting water.*

I learn that the total crop cycle for geoducks is about five to seven years. When they reach maturity, geoducks are harvested hydraulically using low-pressure, high-volume hoses. The machines, which produce water pressure similar to that of a garden hose, have similar hoses that are pushed into the sand next to a resident geoduck. The flow of water loosens up the surrounding sand, enabling a worker to reach in and grab the (now dislodged, and most likely, quite disturbed) clam.

This geoduck's siphons are clearly visible above the sand.

---

* As I watched, an army of mature geoducks haphazardly and randomly squirted streams of water four to eight inches in the air. It reminded me of the "Whack-a-Mole" game at your annual county fair, an irregular and spasmodic (squirting) sequence without logic or pattern.

"The first thing you have to do when you get a geoduck out of the ground is put a rubber band on it," Marco tells me. "That prevents the geoduck from gaping open and losing their internal water and dying. It also increases their shelf life from a few hours out of the water to several days out of the water. We believe the geoduck is less stressed, so they travel better when they have the rubber band on them 'cause they have the same feeling of pressure that they're used to three feet underground."

The day of my visit was a planting day, not a harvesting day. We chat briefly with some of the workers—young men in high boots and equally high spirits—skillfully planting row upon row of geoduck seeds.

> **FAST FACT:** A geoduck planting crew of six workers will try to plant about 10,000 geoduck seeds during a single tidal run.

Stepping over squirting clams, Marco and I head back to the van and the half-hour drive back to company headquarters. Along the way I pepper him with an array of questions about clam aquaculture in general and geoduck farming specifically. I am interested in the number of geoducks the company produced in a single year.

Marco informs me that the company produces about 600,000–700,000 pounds of geoducks a year. He notes that since geoducks are a six-year crop, it takes quite a while to see the returns. There are also a lot of up-front costs; the seeds, for example, are quite expensive, around 50 cents each. However, Marco is quick to point out that the geoduck end of the shellfish business is a relatively small percentage of the company's overall sales.

I'm curious about the markets for geoducks. Marco tells me, "The primary market for geoducks is both domestic and export into Asian markets. Domestically, they are primarily being sold by Asian-American grocery stores. They're being displayed in live tanks and sold live out of the water. Anywhere you go in the world, the geoduck is going to be sold out of an aquarium. They do really well when they're submerged in water."

Banded geoducks for sale in Hong Kong. The prices in HK dollars convert
to about twenty to thirty-two US dollars per geoduck.

I also learn that the other primary market is high-end Chinese restaurants. In those venues, geoducks are generally sold whole, like lobsters.
Each clam will have a price per pound, and diners will select a specific
geoduck from a tank. That clam will be specially prepared and, since it is
quite large, usually shared with an entire family.

I feel like a Food Network audience of one as Marco dishes on some of
the finer points of geoduck haute cuisine:

"There are two cuts of meat on a geoduck. There's the breast or belly
meat, and then there's the siphon or neck meat, and they're very different in
texture and flavor. The siphon is firm and almost crunchy and should not
be cooked; it should be eaten raw or marinated in lime juice. That would

be the most cooking you would ever want to do with it, because it becomes chewy as you overcook it. But eating it raw, it has a wonderful texture.

"The belly meat is much softer and stronger in flavor and can be cooked, you can bread it or fry it, or you can put it in stir fries, or there's a variety of different things you can do with it. It's much more adaptable."*

FAST FACT: Depending on the year and current political situations in various countries, Taylor Shellfish will sell about 50 percent of their geoducks overseas and 50 percent domestically. They distribute approximately 650,000 geoducks annually.

The van cascades through resplendent forests, around pristine bays and inlets, and past long banks of dense foliage. My instructor (and driver) continues my geoduck education as he describes how geoducks are graded.

"Geoducks come in several grades. A #1 geoduck has to be very white in color, nice shell shape, no broken shell or chips, and a nice white color to it. It also has to be between one-and-a-half to two pounds. If it's between one and one-and-a-quarter pounds, it's going to be graded a #1S, which also has a very high price—good quality, just a little smaller.

"Then, you get into #2s. These clams are a little darker, they might have some irregularities in their shape, and they might also be smaller. Number 3s, on the other hand, are going to be a lot uglier geoducks and a lot smaller . . . very small," he tells me.

Marco explains that the pricing goes down as the grades go down. For #1s (the preferred grade), the pricing will be twenty to twenty-five dollars per pound. Thus, a one-and-a-half pound geoduck may sell for about thirty dollars. More often, however, the price is slightly higher—meaning a wholesale price of about thrity-five to thirty-eight dollars per clam. Marco further informs me that a forty-pound box of geoducks might average somewhere in the thousand-dollar range (wholesale, delivered).

I'm curious about how geoducks are used overseas. Marco tells me that geoducks are replacing shark-fin soup as the quintessential wedding food.

---

* For an assortment of geoduck recipes, go to www.geoduckrecipes.com. Included are recipes for fried geoduck, geoduck salads, geoduck soups, raw geoduck, and sautéed geoduck. Be sure to check out the recipe "Geoduck and Avocado Salad with Fruit Salsa." Yum!

In fact, there are large campaigns where young people in China are telling their parents they don't want shark-fin soup at their weddings. As a result, young people getting married in China are often serving geoduck at their weddings.

In short order, we arrive back at company headquarters, park the van, and amble over to the company store. There, Marco dishes up a cardboard cup of piping-hot homemade geoduck chowder for my forthcoming drive to the southern Washington coast. It is a journey filled with exotic flavors, rich aromas, and a culinary spirit unlike any other. I am now satiated both gastronomically and educationally.

Yay, geoducks!

Yay, *Panopea generosa*!

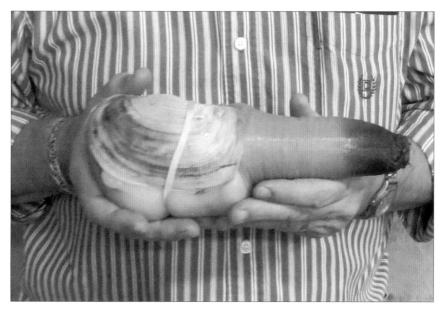

The author's new geoduck friend.

Chapter 10

Clams in Your Back Yard

We are tied to the ocean. And when we go back to the sea . . . we are going back from whence we came.

—John F. Kennedy

A S A KID, MY IMAGINATION WAS CONSTANTLY fueled by a growing collection of adventure-filled books (e.g., *The Hardy Boys, Tarzan of the Apes*) along with a propensity to see almost every cheap science fiction movie ever created in the fifties (e.g., *The Giant Claw, I Married a Monster from Outer Space*). As a result, I made up stories . . . lots of stories. Many of those stories became the essays I would turn into my teachers or the tales I would share with friends as we pedaled through the neighborhood or climbed through the trees in the back yard.

I loved storytelling (and still do) . . . the more imaginative the better!

The ancient people of Hawaii also invented tales and legends about their world. Some of their stories were created to help explain natural phenomenon at a time when scientific knowledge was quite rudimentary. Then, again, some stories have been told so often (and so passionately) that they have been accepted as fact. No doubt several legends have snippets of truth unconsciously and unknowingly embedded throughout the narrative, so much so that it's often difficult to separate actual fact from imaginary fiction.

As the ocean was always an integral part of Hawaiian life, many tales were told of its power and influence. One legend tells of tells of Ku'ula-kai, a man with a human body, but one possessed of supernatural powers (*manu kupua*) that enabled him to direct and control the fish of the sea. According to the legends, it was Ku'ula who constructed the first Hawaiian fishpond—a structure located at Kaiwiopele in Hana on the island of Maui. Ku'ula's pond was located near the shore where the surf breaks, and it is said that he stocked it with a select variety of local fish. He also built a house upon a rocky platform that he called by his own name. It was here that he offered the first fish caught to the fish god, and because of his observances, fish were obedient (*laka loa*) to him; all he had to do was to say the word, and fish would appear.

Another Hawaiian legend focuses on the Alekoko Menehune Fishpond located near the present town of Lihue on the island of Kauai, an historical site that has been on the National Register of Historic Places since 1973. The legend tells how the fish pond was constructed more than 1,000 years ago by the Menehune [pronounced meh-neh-HOO-nee], a race of diminutive and very secretive people who were outstanding engineers and particularly fond of fish. The Alekoko Menehune Fishpond was created exclusively for Hawaiian royalty (*ali'i*) by damming a part of the Huleia River with immense stones used to create expansive walls five feet high and nine hundred feet across. The legend tells of how this pond was constructed in less than twenty-four hours. They managed this amazing task by lining up, side by side from the village of Makaweli, for twenty-five miles, passing stones hand to hand to build the pond.

FAST FACT: Though numerous Menehune legends abound on Kauai, some say the word may have been derived from the Tahitian word *manahune,* meaning commoner, or small in social standing, not in physical size.

While this particular pond (still extant) undoubtedly took more than a day to build (legend), it is clear evidence that the ancient Hawaiians were masters of aquaculture. In fact, archaeological research has revealed as many as four hundred sophisticated fishponds throughout the Hawaiian Islands.

Ancient Hawaiian fishponds (*loko i'a*) were enormous grazing areas where select varieties of fish were kept and raised. In most cases these fishponds involved a shallow area of a reef flat encircled by lava rocks or coral blocks (up to a half ton in weight) constructed out from the shore. These walls were permeable—enabling the ebb and flow of seawater to constantly refresh the ponds and avoid stagnation. At the same time, pond masters often used a sophisticated system of secondary ponds brimming with nutrient-rich water to provide the proper nutrition for their fish "herds." By keeping the fish enclosed, it was quite easy to catch the fish.*

In a culture that honored the earth's abundance, fishponds symbolized the connection Hawaiians forged between themselves, the land (*'aina*), and the gods (*akua*). Shrines at fishponds honored Ku—god of war, fishing, and canoe building—and his wife Hina. Built at the eastern end of the pond, a Ku shrine was often an erect stone symbolizing the rising of the sun, procreation, and the protection of the fish in the pond. A Hina shrine was often placed at the western end, a stone lying flat to symbolize the setting of the sun, growth, and procreation.

Many of those ancient fishponds have been restored in recent years. Ponds have been located on all of the Hawaiian Islands, indicating that this ancient form of aquaculture was not only broad based, but had its roots in a civilization far more sophisticated in food production, specifically aquaculture, than many European countries of the time.

The Alekoko Menehune Fishpond.

---

* In several southern states, "noodling" is a popular pastime. "Noodling" is a common term for hand fishing. This is a method of reaching underwater into natural cavities formed in riverbanks or by tree roots to capture flathead catfish by hand. When a catfish bites on to a noodler's hand, the noodler pulls the fish off the nest and out of the water. Obviously, using a part of your anatomy for fish food is not a sport for everyone. So, if you don't mind, I think I'll just stick to an old-fashioned rod and reel, thank you very much.

Although records are somewhat incomplete regarding the initiation of this system of aquaculture, there is some evidence to suggest that Hawaiian fishponds were fully functional as early as the fourteenth century. By the time Captain James Cook arrived in Hawaii in 1778, Hawaiians were pulling more than 900 tons of fish a year from their respective ponds.

Some of the best-preserved fishponds can be found on the Big Island of Hawaii. Many years ago, when my wife and I first traveled to Hawaii, we made sure the Big Island was on our itinerary. One of the places we were most eager to see was Kaloko-Honokōhau National Historical Park, situated below the volcanic slopes of the majestic Hualālai Volcano. Because of its archaeological and cultural value, this area was designated as a National Historic Landmark in 1962 and was established as a National Historical Park in 1978.

> **FAST FACT:** Besides being the largest island in the United States (4,028 square miles), the Big Island of Hawaii has the southernmost point in the United States and ten of the thirteen climate zones in the world (including, believe it or not, polar tundra). It is also home to Mauna Kea, the tallest mountain in the world (33,100 feet, but only 13,803 feet above sea level). Mauna Kea is taller (from base to peak) than Mt. Everest (a maximum height of 15,260 feet base to peak with an elevation of 29,029 feet above sea level).

As we drove into the park, we were immediately taken by the apparent harshness of the surrounding environment and wondered how this enclave on the Kona side of the island (in the Hawaiian language *kona* means leeward or dry side of the island) would have been hospitable for humans. But, indeed, it had once been a thriving Hawaiian settlement. Nowhere is this clearer than in a visit to the Kaloko Fishpond, an oceanic structure that underscores the impressive engineering skills of the ancient Hawaiians.

Imagine, if you will, a large geographical area (a football field, for example) constructed out of puzzle pieces, each one weighing in at several hundred pounds. The fishpond we saw was like a giant puzzle of interlocking rocks. Stretching for nearly 250 yards, its wall was over six feet high. What was most impressive, however, was not its size but rather the intricacy of its construction. This was a *loko kuapa*, a wall where the stones are dry stacked without the use of mortar. Even more amazing, the stones were not shaped but were arranged with a keen eye to their individual symmetry, size, and position. We both noticed that these walls had been constructed at an angle that deflected and diffused the energy of ocean waves. As such, seawater was able to easily penetrate the porous lava rocks and circulate throughout the interior fishpond. The addition of a sluice gate (*makaha*) enabled water and small fish to enter the fishpond, while preventing larger fish from escaping.

> **FAST FACT:** Hawaii is the only one of the fifty states that does not have a straight line in its state boundary.

As impressive and sophisticated as the fishponds of Hawaii are, they are not the only examples of early aquaculture, nor were they necessarily the first. There is some recent evidence that the earliest aquaculturists may well have been the indigenous Gunditjmara people of Australia. It seems as though the Gunditjmara used the volcanic floodplains near Lake Condah (located in southwest Victoria, Australia) to raise eels as early as 6000 BC. To do so, they developed an elaborate series of dams and channels that entrapped and harbored eels, making them available for consumption throughout the year.*

There are also records documenting the practice of aquaculture in ancient China at least as early as 2500 BC. Because of periodic floods, many river fish became trapped in pockets of water along riverbanks throughout China. Specially trained fishermen kept and maintained these small bodies of water until the trapped fish, typically carp, reached sufficient size for sale (and human consumption). Later, in the fifth century BC, records show that early aquaculturist Fan-Li raised carp in specially constructed ponds, rather than in the previously mentioned (and occasional) river enclosures.

Then, approximately 2,000 years ago, fishmongers throughout China began the practice of selling live fish as a regular feature in town market-places. For the most part, these cultured fish were either maintained in special ponds outside the city limits or were kept in specially woven baskets just inside the marketplace. If you do your grocery shopping in a large supermarket, there may be one or two large tanks of live lobsters being

---

* Eels are not, as you might imagine, a welcome dish in the United States. However, they are quite popular in many cultures around the world, not only because of their ease of preparation, but also for their numerous health benefits. (They are very rich in vitamins, for example.) The Japanese, for example, believe eels to be a culinary cure for lethargy. My own personal cure for lethargy is to run as fast as I can away from the dinner table whenever I'm served a plate of slimy snakelike creatures.

maintained in the seafood section. Thus, your local market may be carrying on a long-standing Chinese tradition.

> **FAST FACT:** As a result of selective genetic mutations over many generations, several varieties of Chinese carp eventually evolved into goldfish. Evidence uncovered in ancient art and literature indicates that the Chinese have bred and domesticated goldfish since at least the Tang Dynasty (AD 618–907).

Early Aquaculture

Let's now shift our geographical and aquacultural attention to North America. Recent anthropological discoveries indicate that humans first entered North America approximately 15,000 years ago. Genetic clues and a reexamination of previous theories (specifically those surrounding the Clovis people in what is now New Mexico) have led scientists to conclude that humans may have been on this continent far longer than previously thought. Yet, what is undeniable in these recent studies is that many of those early immigrants tended to settle in seaside communities, simply (but not exclusively) because of the easy availability of food, both terrestrial and oceanic. (This would be similar, I suspect, to our own propensity to settle in areas surrounded by lots of steak houses and an abundance of seafood restaurants.)

Not surprisingly, the historical and cultural influence of clams on the course of human events has also been documented in North America. One such place is the Strait of Georgia on Canada's west coast. There, rock walls frequently surface in the intertidal zone, the area of shore land exposed during low tide and hidden when the tide is in. What initially appears to be nothing more than random scatterings of rocks are actually carefully constructed "holding pens" used to catch food (e.g., fish) from the sea. One formation, a circular shape almost one hundred feet in diameter, is a clam garden, a flattened area that pools water and creates a most favorable habitat for clams to grow in.

Dana Lepofsky, an archaeologist at Simon Fraser University in Vancouver, British Columbia, believes these gardens and traps, found up and down the Canadian coast, could be more than 2,000 years old. They were most likely used by the indigenous population and serve as artifacts that dispute what the archaeological thinking has claimed was the area's primary staple: salmon.

Close archaeological inspection of nearby kitchen middens suggests that while the red, fatty fish might have been prized, salmon was only available during seasonal runs. Though the early North Americans dried and stored the salmon they caught, it would have taken more than just the usual seasonal catch to feed large (and expanding) ancient communities.

Lepofsky believes that native British Columbians deliberately managed their marine resources for the long term. By combining archaeology with local oral history, she and others are postulating that these societies oversaw an entire oceanfront ecosystem, one offering a diverse bounty of marine life, including fish and, of course, clams. As other sites are revealed through shifting tides and geological events, it is assumed that these theories about early aquaculture in North America will receive additional confirmation.

> **FAST FACT:** In British Columbia, the oldest known archaeological sites point to continuous human occupation on the coast for at least 11,000 years.

Although the archaeological record of this early period is still evolving, these sites may hold some of the richest information about the first people who arrived on North America's west coast. Undoubtedly, these are the same bands of migratory people who traversed the land bridge between northeastern Asia and North America and made their way down the coastline of what is now Alaska and western Canada and into more temperate regions of the North American continent.

Ever since those first Native Americans discovered clams as a convenient source of food, it has been harvested along the coasts of the United States with gusto by successive generations. Readily available along both coasts, clams are easy to harvest requiring a minimum of skill and equipment.

From those early successes with marine aquaculture, you might expect that Americans would embrace this "farming" technique with gusto. Unfortunately, the science and practice of aquaculture languished—perhaps because it was seen as a localized practice or because of the primitiveness of the enterprise.

As a result, it wasn't until 1859 that we saw the first recorded instance of formal aquaculture in the United States. That was when Stephen Ainsworth of West Bloomfield, New York, began experimenting with the cultivation of brook trout. Ainsworth was not seeking to establish a commercial venture, but was more interested in aquaculture as a hobby. Then, in 1864,

THE SECRET LIFE OF CLAMS

Seth Green, in nearby Caledonia Springs, New York, set up and began running a highly profitable fish hatchery. Green's profits came primarily from supplying fish eggs to numerous hobbyists interested in cultivating fish for their own commercial and recreational interests. Shortly after, in 1889, Adolph Nielson built a major fish hatchery in Newfoundland, Canada. This hatchery took advantage of several recent technological advancements in commercial marine aquaculture.

However, from the late 1800s on through most of the twentieth century, formal aquaculture never really took hold, most likely because of a lack of financing or commercial appeal. It wasn't until the latter part of the twentieth century that aquaculture was embraced as a profitable economic venture (most certainly because of threatened natural resources, health concerns regarding red meat diets, and an increasing national demand for seafood). Since then, it has taken off like a rocket.

Today, aquaculture—specifically clam aquaculture—is a commercial bonanza. For example, Florida aquaculture sales (in 2005) totaled seventy-five million dollars. (About 70 percent, or fifty-two million dollars, was directly attributable to clams.) Today, throughout the Sunshine State, there are as many as twenty-five hundred people employed in the harvesting, processing, and distribution of shellfish.* According to the California Aquaculture Association, aquaculture creates a two-hundred-million-dollar annual contribution to the California economy.

Virginia is another state that has taken advantage of the profitability of clams. According to 2013 figures from Virginia Sea Grant, shellfish aquaculture has contributed more than eighty-one million dollars each year to that state alone. In Washington State, the total value of aquaculture in the state is approximately ninety-seven million dollars. In addition, there are approximately two hundred shellfish farms in Washington with clam farms contributing more than seventeen million dollars annually to the state's economy. Suffice it to say, the growth of the commercial aquaculture industry has not only contributed mightily to the economic resources of select coastal states, but also to the overall preservation of sufficient stocks of clams for a growing population of seafood consumers.

---

* In 2005 Florida aquaculture took up 3,010 water acres split up between 359 farms.

Imagine you are an aquaculturist—specifically a clam aquaculturist. I'm sure you wouldn't be surprised at some of the obvious advantages of farming clams versus farming fish (or other kinds of animals). When farming clams:

- Your product doesn't require expensive or expansive cages to keep them contained.
- Your product is not prone to escape out into the wild.
- Your product pretty much stays in one place for most of its life.
- Your product doesn't require enormous quantities of expensive food.
- Your product is not labor intensive.
- Your product is fairly easy to harvest.
- Your product is, for the most part, well behaved.

Clam Aquaculture Up Close

"In some ways aquaculture has this kind of negativity, and maybe rightly so, because there's been some problems with farmed salmon; there's been some problems with escapees and with the food that they eat. But for clams, for Cedar Key clams, it's very, very simple. They're eating in a pristine environment. There's no industry around us, the waters are very clean as far as pollutants, and the waters are also very rich with algae. These guys are living like a wild clam would live, and they're growing like a wild clam would grow."

Jon Gill is one of a select legion of seafarers—a fraternity of watermen and waterwomen who have sought their fortunes in and around Cedar Key, Florida, a cartographic speck some fifty miles southwest of Gainesville. Located on a group of small islands, Cedar Key (named for the Eastern Red Cedar, *Juniperus virginiana,* which once grew abundantly in the area) is a working waterfront community jutting three miles out into the languid waters of the Gulf of Mexico. This tiny coastal enclave has a total population of just under nine hundred full-time residents. But don't let its small size deceive you; this is a town with plenty of charm to go around. (In 2011, Cedar Key was voted one of the "Ten Coolest Small Towns in America" in Budget Travel's annual survey.)

FAST FACT: Cedar Key is also one of the oldest bird and wildlife refuges in the United States.

Welcome to Cedar Key, clam hotspot.

Travel southwest of Gainesville, Florida, on Route 124, cross four bridges, and at the end of the road you'll discover an out-of-the-way town with no fast-food franchises, no soaring skyscrapers, no hustle and bustle, and absolutely no traffic lights. Most of the businesses are mom-and-pop operations housed in well-worn storefronts. Victorian buildings, traditional cracker woodframe homes, long piers stretching out into the Bay, spritely art galleries, placid waters dotted with kayaks, outstanding seafood restaurants (you must have a bowl of clam chowder at Tony's Seafood Restaurant!), brilliant sunsets, and a slower pace of life are the norm here. (Look for all the "snowbirds" tooling around town in rented golf carts.) This is old-time Florida, a place where folks on the sidewalks aren't visually glued to their smart phones or permanently fixated on their iPads. These are down-home, friendly people who actually take the time to greet you on the street and welcome your presence.

For most of its history, Cedar Key was a sleepy fishing village just south of the mouth of the Suwanee River. Surrounded by long acres of protected sanctuaries, mesmerizing salt marshes, panoramic tidal creeks and numerous barrier islands rich with wildlife, Cedar Key is an environmental paradise. However, Cedar Key's existence changed dramatically when Florida

voters passed a statewide gill-net ban in 1994. The ban effectively elimi-
nated the harvesting of mullet and other marine fish, directly affecting the
island's numerous fishing families by separating them from a livelihood
practiced by generations of fathers and sons, and mothers and daughters.

But, instead of just enacting a new regulation, the state also stepped in
and provided job retraining programs for those unceremoniously divorced
from their life's work. For many, this was a transition to shellfish aquaculture, specifically clam farming, as an alternative employment opportunity.
Now, more than thirteen hundred acres of state-owned submerged lands are
dedicated to aquaculture leases. Since the mid-nineties, clam farming has
added an estimated forty-five million dollars a year into the area's economy.
Just as important, it supports more than five hundred jobs—like Jon Gill's.

Jon, along with his brother-in-law, Shawn Stephenson, is co-owner of
Southern Cross Sea Farms, a rustic assortment of weather-beaten buildings
just off the side of State Route 24 and just before you arrive in the town of
Cedar Key. Jon is tall and narrow, with a shock of brown hair and languorous
eyes. He reminds you of a star high school basketball player whose playing
days are long over, but one who still commands attention and respect.

It was a brilliant Friday when I first met Jon in the second-floor office
he shares with Shawn—an office that looks like the aftermath of a small
hurricane with piles of papers, books, charts, maps, and assorted paraphernalia scattered and stacked and stuffed into every nook and corner of the
cramped space. In short, just like my office.

After a quick conversation, Jon takes me down the rickety outside stairs
where we join up with two couples (fellow senior citizens) ready to take the
one o'clock tour of Southern Cross's facilities. Our first stop is at the clam
breeding table. As Jon points out, it is here that adult clams are induced
to spawn by the artificial manipulation of water temperatures. The process
begins when a bed of clams* is placed on the table. However, Jon is quick
to add, "There is no way to tell if they are males or females—they all look
exactly the same until they start the spawning process."

Jon then proceeds to give the five of us a short lesson in sex education—
bivalve style. He points out that there may be up to two hundred clams on
the table—approximately one hundred females and one hundred males.
Workers begin to adjust the water temperature on the breeding table until,
eventually, a single male gets things started off. Or, as Jon puts it, "Little

---

* A group of clams is known as a bed. Other animal group names include a murder of
crows, a crash of rhinos, a parliament of owls, a bask of crocodiles, a smack of jellyfish, a
gaze of raccoons, a streak of tigers, and, my favorite—a wisdom of wombats!

guy sitting here, he says, 'The water temperature feels right, I think I'm just gonna let it go.' So he lets off his sperm. His sperm, what he lets go, is milky, silky. The female looks exactly like the male; you wouldn't know the difference. However, the female has a more [particular] look and her eggs come out a little bit different. You are talking microscopic, little baby eggs that are very hard to see. The male would go and then that sperm would go all over, it would be circulating throughout, all over the table. The girls would say, 'Well if I want my eggs fertilized, this is the right time to do it.'" Jon is on a biological roll, and he informs the five of us that this entire process may take anywhere between two and four hours.

Jon Gill at the clam table.

Jon is emphatic when he tells us that the clams get absolutely no enjoyment and no pleasure out of this. As Jon puts it, "all they are doing is letting it go."

One of the women in the group is starting to get a little nervous and begins fumbling through her purse for her cell phone to check for any messages. ("Please, oh please, someone send me something . . . anything!")

But, like a biology professor on steroids, Jon presses on: "A girl would start to go, we would recognize that it's a girl because it's more particulate, more sandy looking as it's coming out. We would grab her, we would not let her spawn on this table, we would put her in one of these little tanks with about that much water and keep her released eggs in there. The reason being, we have these one hundred males and one hundred females; yet, it'd take only about two or three of those males to fertilize one hundred females. If you let them all go at the same time, there is actually enough sperm that would kill the eggs. They would penetrate so many times it would kill the eggs. We don't want that, so we monitor how much sperm we put in there by taking a few of the pretty boys out, putting them in a little cup, letting them spawn there and then monitor it via the microscope to make sure we have the right quantity of sperm-to-egg ratio. It's not perfection, it's just that one girl could have two million eggs and one guy could have close to a billion sperm. The penetration happens very quickly—within an hour they are fertilized eggs."

("Please, oh please, someone send me something . . . anything!")

With images of an all-out clam orgy dancing in our heads, Jon slowly moves the group over to a series of six-hundred-gallon tanks, each of which can hold between six and ten million larvae. Here the tiny larvae spend the next few days of their life in filtered, sterilized seawater where they get increasingly and progressively larger. While in the tanks, they are fed a continuous diet of cultured phytoplankton during the critical ten- to fourteen-day larval phase. Then, after about two weeks, the larvae begin to settle out of the water column, or metamorphose. In essence, they lose their ability to swim.

Jon tells us that, for this phase of clam life, several species of phytoplankton are purchased from nationally certified laboratories. He later explains that Southern Cross has tried to grow their own phytoplankton, "but it really requires a lab to segregate the one particular type of algae that we would need and then to make that into an actual body of it instead of it being a single-celled organism. We don't have the ability to do that," he explains.

FAST FACT: Cultured phytoplankton is very expensive. I found the following online prices: $21.97 for a sixteen-ounce bottle from one supplier, and $101.99 for a one-hundred-milliliter bottle from another. However, if you would like to try growing your own batch of phytoplankton at home, you might be interested in a complete set of directions (including necessary materials) at www.melevsreef.com/phytoplankton.html.

As it turns out, the commercial algae is the most nutritious food available for clams. Although it begins its development in small test tubes (as delivered by the labs), it is eventually transferred into a massive array of oversize translucent tanks at Southern Cross, where it multiplies and flourishes. Fluorescent lights are mounted behind these immense containers to create a perfect growing environment for this vital clam food.

Large tanks of specially grown phytoplankton.

As Jon explains, the food (algae) is eventually mixed with filtered water. Each addition is constantly monitored, and, when the time is right, the algae is transferred into a specially designed greenhouse. We learn that Southern Cross grows various types of phytoplankton at different levels of maturity so that there is always a steady supply of food for the clams.

After metamorphosis, the baby clams fall to the bottom of the large tanks. When they do that, the tanks are drained, and the clams are sieved and housed in the hatchery, basically in sixty-gallon tanks. They live in these tanks (with screen mesh on the bottom), where they get fed and cleaned every day. Jon also points out that these seeds are microscopic and, thus, are vulnerable to fluctuating environmental conditions. As a result, they are carefully maintained in the hatchery for another thirty to sixty days until they reach approximately one millimeter in size (about the diameter of a pinhead).

Jon moves the group into another room with an array of large flat tanks. This is the nursery. Essentially, the nursery is a protected area that serves as an intermediate step in the clam development process. This is a place where the seed clams have a new food supply (wild phytoplankton from the Gulf of Mexico) and continuous protection from any potential predators.

Like a practiced kindergarten teacher, Jon shepherds the group outside and over to the dock area, an expanse just offshore with large floating tanks in the water around Cedar Key. The dock system consists of upweller tanks that deliver large volumes of water to the clams. "[Here] they're eating natural algae," Jon explains. It is here that the clams are raised in a holding system for one to three months (depending on the time of year). Once they reach four millimeters (about the width of a CD case), they are taken and placed in clam bags or nursery bags with very fine mesh. The nursery bags are stocked with about fifteen thousand seed clams per bag. Those bags are then planted out on the clam leases,* where they will remain for about two to three months. During that period, they will grow to about the size of a dime.

Jon reaches into one of the upwellers and takes out a handful of teeny-tiny clams to show us. I estimate he has approximately one to two hundred clams in his hand, each one the size of a very small aspirin tablet. As I snap off a couple of photos, it is clear that these miniscule critters are destined for a cushy life—one with unlimited supplies of algae to consume and unlimited protection from bipedal predators—at least until they wind up two years later on someone's dinner plate in Miami, Los Angeles, or Chicago.

---

* At any one time, there may be twenty to thirty thousand bags planted out on the clam leases.

Teeny-tiny seed clams.

"After the clams reach approximately 12 millimeters in size, they are brought back in and transferred into grow-out bags. Each grow-out bag holds about twelve hundred seed clams, and these will spend about a year to eighteen months on our fertile clam leases three miles offshore of Cedar Key," Jon tells us.

The grow-out bags are made of a polyester mesh material. Once boats arrive at the clam leases, clammers will stake the bags to the bottom, planting them in rows much like terrestrial crops. After several weeks the naturally occurring sediments from tidal and wind actions enable the clams to bury in the bottom. There, the clams ingest the rich Gulf-water algae and (like your average American teenager) they will grow, grow, grow. After twelve to eighteen months, the grow-out bags are winched up from the bottom and loaded onto boats. The boats return to the dock and unload their catches.

Jon moves us over to a series of tumblers—large revolving contraptions that roll and tumble the clams to get rid of sand, shells, and anything that's not a live clam. The tumblers also rinse and clean the clams for processing. It is a cacophony of sounds, as several tumblers are all in operation at the same time.

As we move back into the main building, the group weaves its way through an array of large plastic baskets haphazardly arranged across the wooden loading dock. These are the newly tumbled clams awaiting their entry into the processing room.

Clams a waiting entry to the processing room.

As Jon takes us into the chilled processing room, the noise level escalates, the activity intensifies, and conversation becomes quite difficult. As Jon points to one of the machines, we watch as baskets of clams are poured into one end, tumbled down a series of sizing rollers, sorted (into four different categories) according to size, and eventually shuffled down into color-coded mesh bags.

Jon informs us that a clam is measured into one of four categories: little necks, middle necks, top necks, and pasta necks. Although there is variation in the sizes, the clams are all the same species. The company's graders are equipped to handle those four categories (or four different sizes), as measured by the width of the clam. The largest size is an inch and an eighth, the next one down is an inch, the next one is seven-eighths of an inch, and the final or smallest one is three-quarters of an inch. Jon tells us that the clams are put in various size bags according to what a particular customer (for example, a specific restaurant) wants. Those bags may range from small (12 count or 50 count) to very large (240 count), but most customers request the mid range bags (100 count or 200 count).

After the clams are bagged, they are shipped out by truck and airplane to customers all over the United States. As Jon points out, clams can be harvested, processed, and shipped to customers all on the same day.

> **FAST FACT:** Southern Cross Sea Farms ships approximately thirty million clams each year.

After the tour, Jon and I retire to his office. While he ingests a quick lunch, I pepper him with questions. I quickly learn that Jon has been a life-long fisherman—from high school to the present. After the state's ban on net fishing, he was retrained in clam farming, something he's been doing for the past seventeen years. Then, in 2009, he and his business partner, Sean (also his brother-in-law), purchased Southern Cross Sea Farms, and he's been involved in all the details of the operation ever since.

"We're a vertically integrated business. What that means is that there are lots of companies that just farm clams and there are lots of companies that just process clams. But Southern Cross starts the clams from scratch. We have a hatchery, we have a nursery, we have our field planting, our farm, the processing plant, and we have the shipping. It's the full cycle of the clam business—that's what vertically integrated means."

Sean enters the office and explains to me that the company, besides being vertically integrated, is also trying to branch out into some new directions. That is, they are also experimenting with growing oysters and other species of shellfish. In so doing, they are working on building new markets throughout the country in addition to distributing clams to as many places as possible. Jon provides the statistical data when he informs me that Southern Cross sells about 25 percent of their clam seeds to other shellfish farmers, while they plant the other 75 percent.

In response to my question about some of the challenges of clam farming, Jon tells me, "The good part about clams is that we never have to feed them after they leave the hatchery. In most fish farming—shrimp farming, catfish farming—where they're farming fish, they have to feed them. They have to buy processed food; whereas our clams live out in the wild just like clams everywhere do." Jon also points out that clam farmers keep their clams in one spot and constantly protect them. Since clams are naturally filter feeders, they are always cleaning up the immediate environment. As a result, clam farming is one of the most environmentally friendly occupations around.

> **FAST FACT:** A single clam can filter between five and ten gallons of seawater a day, retaining particles as small as two microns. As a result, a two-acre clam farm can filter approximately five to ten million gallons of seawater each day.

During the tour I had noted that Southern Cross was definitely a low-tech operation. I wanted to get Jon's "spin" on that part of the business, and he explained that they are currently not into all the technological innovations. "There's probably a lot of room for growth there," he explains. "Our most technological part of the business might be our grading machines, but everything else is more manual-type labor." Jon clarifies by stating that they would like to see some technological changes in the future—for example, an automated processing system that doesn't require so much labor. However, they've discovered, through trial and error, that clam farming is a relatively straightforward process. Clams, Jon points out, are relatively simple organisms that don't "require a ton of work. We can keep the price very low by running a lot of them through here, continually doing what we've always been doing."

Jon reminds me that at the time of the net ban (1994) there was absolutely no clam industry in Cedar Key. "There was none of the three hundred or four hundred leases that are here and the three to four hundred people that are employed, right here locally, by the clam industry." Clams, according to Jon, have been very good for the local economy. But the clam industry, like all newfound industries, is going through continual transformations, adjustments, and changes. In Cedar Key, the industry started off with "basically three hundred different farms, and now it's down to about fifty farmers with ten wholesalers that are big farmers, too."

However, as Jon is quick to point out, there's still plenty of room for growth. He makes it clear that Cedar Key is primed to grow and produce many more clams than it already does. Jon is passionate when he tells me there's room, both economically and environmentally, for additional farmers to claim their share of the economic pie or for current clam farmers to expand their operations without a serious or deleterious effect on the local environs. "It's kind of a nice industry to be in, because farming's not totally hated yet, like commercial fishing was. [Clam] farming still has a fairly decent name and there's [room] to grow. People like to eat clams, so we want to keep continually producing as many as we can."

I wanted to know if the current market could support the kind of growth and expansion Jon believes is possible in this little slice of paradise.

"The clam is a very cheap seafood, it's still not expensive," he points out. "You go buy king crab, and you will pay through the nose, but a clam is still very cheap. We just probably need to start with some new markets and beat off some of the foreign competitors who market their products here."

A conversation with Jon is like a conversation with a diehard evangelist. It quickly becomes apparent that there is both passion and commitment involved in any successful clam operation such as Southern Cross. That Jon and his business partner have those traits is clearly evident; that they are at the forefront of a dynamic and exciting business venture is equally apparent. Jon puts a final emphatic punctuation mark on my visit when he tells me, "It's a lot of work around here. I work with my brother-in-law, my sister, and my wife. My kids work here a lot. We're just a hard-working business . . . and we're always growing."

As I make my way down the rickety wooden steps on the outside of the building, I am struck by the fact that clam farming, as practiced by Southern Cross, is a balanced economic and environmental venture—one that gives back as much as it takes. It's as much a stewardship of the natural world as it is an occupational choice. It's ecological replenishment in concert with company profit. And it's biological sustainability wrapped around long-term growth. A damn good business model, I'd say!

Life of a Professional Clammer

"This is the best job I've ever had. I get to watch the sun come up and the sun go down. I get to see all kinds of sea life—crabs, sharks, sea gulls, catfish, and dolphins at play. It's absolutely glorious out there; there's a surprise every day."

For Todd Smith, the life of a professional clammer is the best of all worlds. Although he's only been doing it for a short time, he has fallen in love with every aspect of the job. It is his raison d'etre, his salvation, and his true calling in life. Working for one of the shellfish wholesalers in Cedar Key (not Southern Cross) has given Todd purpose and perspective. Or, as he put it to me, "Like life, it's always an adventure."

Standing at just under six feet, Todd's thin physique belies a well-developed musculature—biceps strengthened and worked by daily hefting of 125-pound bags of clams from the water to the boat and from the boat to the loading dock. With dancing blue eyes and a long goatee, Todd radiates exuberance as well as a passion and motivation for hard work—all necessary ingredients for successful clammers.

For most professional clammers, the days are very long and arduous. It's not unusual for Todd to leave his house at six o'clock in the morning and then return home at six at night. In busy seasons (like the weeks just before the Christmas holidays) he frequently doesn't get home until nine or ten o'clock at night. Then, it's back out early the next morning.

Todd's workday begins soon after he arrives at the dock. There, he and his coworkers quickly don their wet suits and head out into the Gulf on clamming boats, often when the sun is just beginning to peek up over the horizon. They anchor their boats at an offshore "farm"* to haul in bags of clams that have been sitting in the rich Gulf waters developing and growing from tiny critters to marketable delicacies (a process that takes up to eighteen months). They may also use this time to plant new clam seeds (in bags) that will be harvested in successive years. The number of bags harvested on any one day will depend on the number of wholesale orders the company has received from restaurants and seafood purveyors all over the country.

> **FAST FACT:** At harvest, each clam bag can hold between five hundred and one thousand marketable clams. Each bag will bring in around fifty to one hundred dollars.

For you and me, most of our occupational challenges may revolve around constant deadlines ("I want this done by noon, yesterday!"), frequent communiqués between colleagues ("Did you see what she wore to work today? Unbelievable!"), endless directives from our superiors ("Of course I expect you to be here on Saturday!"), and a never-shrinking, always-expanding "To Do" list. Todd tells me that most of his occupational challenges involve the weather and the tides. "You never know what you're going to run into," he says. "The boat may swing around, the anchor may come loose, the rain may be beating down, the nets may get tangled, or the tides may be working against you. You have to be strong to be a good clammer. You have to like the water and be a good swimmer. You also have to be able to think fast. Most important, you have to be dedicated . . . you have to be highly motivated."

---

* There are currently about 170 grow-out businesses located off the coast of Levy and Dixie Counties in Florida.

For most clammers in Gulf Coast waters, the greatest challenge of the job may not be the constant hefting of clams bags but something slightly more sinister—stingrays. As Todd puts it, "Getting stung by a sting ray is a rite of passage for most clammers." He's quick to mention that, although he hasn't been stung anytime in the past three years, he has heard that it is the worst pain possible (often treated on the spot with snakebite kits).*

After Todd and his coworkers load the boats out at the company's aquaculture lease, they travel back to the dock in Cedar Key. There, they unload the bags at the processing facility, where the clams are graded, counted, weighed, bagged, and placed on pallets. Then, the clams are loaded into refrigerated trucks that will take them to the airport to be shipped out and delivered to buyers—usually within twenty-four hours of their exit from the pristine waters around Cedar Key. There are currently about twenty wholesalers in the area (like the one Todd works for) who process between 150 and 200 million clams annually.

When I ask Todd what he loves most about his job, his eyes sparkle and he tells me he loves everything: "I love being out on the water. I enjoy working with the other clammers, as they have become part of my family. I love the physical and mental aspects of the business as it keeps me in shape in so many different ways. I find the whole business of clamming extremely interesting."

As I have quickly discovered in my travels around the country, clammers are a unique breed. Todd cements that appellation when I pose a final question: What is so special about clamming or about clammers? Again, with his characteristic enthusiasm he responds, "Clammers are always interesting people . . . wonderful people! Most of them used to be fishermen, so they have a passion for the sea. They're hard working, smart and friendly. They are big on giving people second chances, and they'll open their arms for everybody. Best of all, it's a solid, honest living—and the money is good!"

So, dear reader, what do you say? Are you ready to go clamming with me tomorrow morning?

---

* Biologists agree that the following six creatures produce some of the most powerful and deadliest stings of any in the animal kingdom: sea wasp, cone shell mollusk, stonefish, sting ray, box jellyfish, and bullet ant. Note how many of those critters live in the ocean. Four words: Be careful out there!

Chapter 11

Low Tide, Clam Rakes, and Education

Only when the tide goes out do you discover who's been swimming naked.

—Warren Buffet

JUST LIKE 74,059,000 OTHER AMERICANS, MY WIFE AND I own a cat. (Well, to be more precise—we are owned by a cat.) Our feline's name is Tubby, and truth be told, he comes by that name quite honestly. He was the chubbiest, plumpest, and heaviest cat in his litter of eight—the one who always pushed his siblings out of the way to get first dibs on one of his mother's teats. It was clearly evident that, from a very early age, he knew how to throw his weight around—a trait he maintains to this day. In addition to his daily cup of cat food, he often lurks around the pantry waiting for someone to leave the door ajar, dips his paws into any bowl or pot left in the kitchen sink, or perches in a prominent location to observe any and all dinnertime meals. (Trust me, having a cat watch you eat dinner often reminds me of several scenes in Alfred Hitchcock's absolutely creepy movie, *The Birds* [1963]. I never know when, or if, I'm going to be pounced upon.)

Now Tubby is a certified senior citizen. In human years, he is twelve years old, but in cat years, he is sixty-five. It should be noted, however, that his advancing age has not hindered him one whit—he is as lithe as an Olympic gymnast and as speedy as a world-class sprinter, especially when the prospect of food is involved.

It has been said that cats are often territorial, and Tubby certainly fits that descriptor. He has taken to patrolling, not only our four acres of mountainous terrain, but the acreages of at least three of our far-flung neighbors. He is as protective of this hummocked territory as would be any medieval ruler of a feudal domain.

He is king of the mountain, so to speak.

We have, however, had to restrict his wanderings, particularly in the spring. That's because he enjoys nothing better than chasing various winged creatures during his outside forays. He never captures any; for him, it's the thrill of the hunt. He particularly enjoys running beneath birds flitting from tree to tree, or those twittering on overhanging branches. It seems he has a particular fondness for colorful birds; the robins and blue jays who swoop from fence post to fence post are some of his favorite "playmates." Although he enjoys dashing and darting with these winged creatures, I suspect he would have his hands (paws?) full should he ever come up against the Raven.

Especially, the Raven!

As you'll recall from chapter 3, the Raven represents many things in Native American lore—a symbol of change, a creature of metamorphosis, a healer, and a keeper of secrets. Many tribes honor the Raven as a cultural hero. He is a benevolent transformer figure who helps people and shapes their world. Yet, at the same time, he is also a trickster character, and many stories about the Raven center around his frivolous or poorly though-out behavior—behavior that often causes trouble for him and the people around him. Above all, the Raven is the Native American bearer of magic as well as a harbinger of messages from the cosmos. (Message to Tubby: Don't mess with me!)

The following legend from the Pacific Northwest, retold from Mary Mahoney's "Raven and the First People," tells of Raven's discovery of clams on the beach . . . and how we came to be.

> Raven was bored. He was walking along the beach, looking for
> some new way to amuse himself. As he walked along the beach,

the blue ocean in front of him and the green forest behind him didn't seem interesting. Raven wanted to play, but there was no one to play with.

Then he heard a strange sound, unlike any sound he knew. He looked up and down the beach. Where was it coming from? As he walked he noticed a large white clamshell lying in the sand. Inside the clamshell were tiny creatures, unlike any he had seen before.

Raven bent down to get a closer look. The creatures seemed afraid of him, so he began to coax them in a gentle voice, "Come out. Come out. Don't be afraid. I won't hurt you."

A few of the creatures came out of the clamshell. They were very different from Raven. They had no feathers, no wings, and no beaks. Like him, they walked on two legs, but they had arms, faces with mouths, and black hair. They spoke to each other in a language that Raven didn't understand. These tiny creatures were the first humans.

Raven enjoyed watching these humans play and explore the world. After a while, when he was beginning to feel bored again, he noticed that these humans were only men. There were no women. Raven had an idea. He wondered if he could find some women. He searched for a long time. Then he saw some clams. He opened one of the clams and found some lovely, tiny women. He brought the women to the men.

Raven enjoyed watching the behavior of the men and women. He saw them begin to pair off and have children. The human families moved to other parts of the island. Since that day, many generations of humans have grown and flourished, and Raven has never been bored.

Clamming Equals More Than Food

Humans plus clams plus beaches have been an irresistible combination for centuries, if not millennia. That you and I are drawn to beaches and that clams live on those beaches is one of the eternal constants of nature—an immutable pairing that celebrates both culture and gastronomy. For many people, the opportunity to gather one's dinner along the shore is a way to reconnect with nature as well as reconnect with a basic primal instinct.

Clamming is for everyone! These clammers are using clam tubes to pull up razor clams.

Many folks enjoy clamming simply because of the opportunity to get outdoors. Others enjoy the sport of clamming because of its simplicity: it is hunting at its most basic level (without the need for noisy, long-barreled armaments, explosive projectiles, bright orange outerwear, and funny knit-ted hats). Still others are drawn to this traditional summertime activity because of the many opportunities to "chill," that is, to get away from the everyday "rat race" and be one with the larger world. The fact that amateur clammers often get to revel in a side of nature not often experienced (or appreciated) is also a contributing factor.

Clamming is also a way of reconnecting with friends and family mem-bers—a way to bring back pleasant childhood memories or to forge new memories with parents, grandparents, and children. In fact, there is no age limit to being a clammer, and everyone in the family can participate in gathering a meal, enjoying each other's company, and establishing critical bonds of communication.

> **FAST FACT:** The Washington Department of Fish and Wildlife reports, "It is not unusual to have as many as a thousand [clammers] per mile [of beach] during a nice spring weekend day."

For me, clamming is much more than an activity—it is an experience! It is an opportunity to stand at the edge of the ocean and to reconnect with the past—with those whose only method of survival was to embrace the bounty of nature. Clamming also brings a sense of tranquility—a way to slough off the burdens and anxieties of our "mile-a-minute" life and enjoy natural beauty. When clamming, I know I'm doing something that has been around for millennia and something that will continue to provide pleasure for many more millennia. Clamming, for me, is all about bonding—bonding with nature and bonding with those we love. It may well be the perfect family activity.

Water In, Water Out

However, before you can be an official clammer, you need to know a little science—specifically oceanography. Because clams are typically gathered at low tide, it is important that you know something about tides and how they work.

Simply put, tides are the natural rise and fall of water along the ocean's shore caused by the pull of the moon's gravity on the earth as it spins around on its axis once every twenty-four hours. When the moon's gravity pulls seawater toward land, the water level rises—this is high tide, when a large section of the shoreline is covered by water. When the moon pulls water away from land, the sea level falls—this is low tide, when much of the shoreline is exposed to the air. In general, low tide occurs approximately twelve hours and forty minutes after the preceding low tide. By the same token, high tide occurs at this same interval. That means that during a twenty-four hour day there will be two high tides and two low tides.

Are you still with me?

Okay, here's another way to look at it: low tide always follows high tide by just over six hours, and the time of each advance is about forty minutes each day. So, if low tide is at 10:00 p.m. on Saturday, it will be at about 10:40 p.m. on Sunday. Oceanographers refer to this pattern as "tidal oscillation", a pattern that is completely predictable, making it relatively easy to know exactly when low tide will occur or when high tide will occur on any beach on any day.

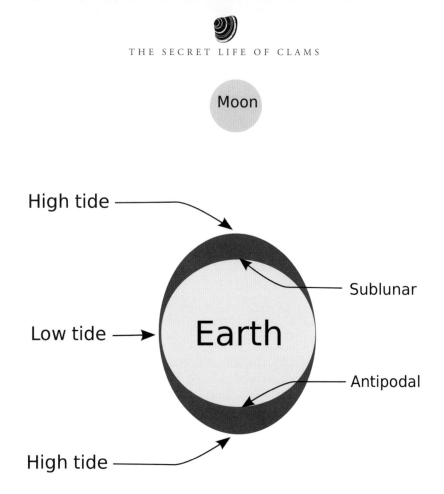

This diagram shows how the moon's gravitational pull affects the tides.

Tides

Many newspapers list the tide times for the day or week. Tide books are also sold at fishing tackle stores, bait shops, boat landings, marinas, and marine aquariums. You can also obtain tidal information for both the East and West Coast from the National Ocean Service of the National Oceanic and Atmospheric Administration (NOAA). The following information is available on its website (oceanservice.noaa.gov):

- **Tide Predictions**: Generate a graphical display or a tabular listing of daily high and low tide predictions for more than three thousand

locations around the nation. Predictions may be generated up to two years in advance.
- **Real-Time Tide Data**: Access current water levels from more than three thousand tidal stations.
- **Historic Tide Data**: For a given NOAA tide station, retrieve historic tide data from the earliest to the most recent dates for which data is available.
- **Tides Online**: Choose a tide station by state and location to view current tidal information, wind speeds, air pressure, and air temperature.
- **Tide Station Index**: Generate a per-state list of all NOAA tide stations, including station number, name, location, installation date, and more.

Tides occur along that portion of the beach known as the intertidal zone. By definition, the intertidal zone is the selected area of the shoreline exposed or covered by the tides. The intertidal zone extends from the highest wave-slashed rocks on top down to areas of the shore that are exposed only by extremely low tides. Different plants and animals live up and down the intertidal zone depending on how well they can tolerate exposure to the air and the crashing of the waves, or can find appropriate food sources there. Organisms that need to be underwater a lot (fish, seaweed, crabs) live near the bottom of the intertidal zone. Other organisms that can tolerate both wet and dry conditions (barnacles, algae, limpets) live near the top of the intertidal zone.

Got that? Okay, now it's time to refocus on clams!

As you'll recall, clams are primarily underwater creatures, establishing their residence, usually one to nine inches deep, in the shoreline sand or mud. Depending on the species, they will be particularly abundant in the area closest to the tide line, or that region under water the most during the endless succession of tidal changes. Since they are filter feeders, clams spend the majority of their underwater time feeding. During those times when the tide is out (low tide) and they cannot bring the water down to them, clams are typically stationary. This is the time best suited for clamming.

FAST FACT: "When the tide is out, the table is set."—Tlingit proverb

Oh, one more thing about those tides—tidal action is constant. That means that the tides are either going out or coming in. But what is most interesting is that the slowest movement of the water takes place in the last twenty minutes the tide is going out *and* the first twenty minutes the tide is coming in. This forty-minute slice of time is known as slack tide, a stretch of time that brackets each of the two low tides of the day. This is the absolute best time to hunt for clams simply because the area in which they are most abundant—the low-tide mark—will be revealed to any and all potential clammers.

Signs of the Times

"But the clams are under the sand," you say. "How do you find them if you can't see them?" you may also ask. Actually, it's quite easy, because clams often leave some clues as to where they are hidden. Some of the most noticeable clues are small holes in the sand—holes in which their elongated siphons are often positioned. By walking along the low tide mark you can often see small holes of various sizes. Occasionally one or more of the holes will eject a small stream of water (as you may recall from chapter 5, some clams are called "pissers," because they often squirt water from their siphons), an indication that a clam is aware of your presence. In short, your footfalls have given you away as a potential predator—a predator the clam would like to avoid at all possible costs.

Another sign that clams are present beneath the sand is the presence of clam scat on the surface of the beach—frequently around the siphon holes. These tiny threads are approximately one-eighth-inch long, look like very, very tiny caterpillars, and are a sure sign that a living (soon to be deep-fried) clam is close by.*

Depending on the beach, some of the best clamming areas are those covered by water all the time. These may include shallow bays, long stretches of the intertidal zone, and sheltered inlets along the shore (the New Jersey shore is famous for these). Since the clams are completely under water there are no visible signs of their presence. In those cases you will discover that

---

* Among nature's most distinctive animals are the caterpillars of the skipper butterfly family. Because slow-moving caterpillars are a favorite prey for birds and other meat-eating insects, skipper caterpillars have a defensive ability that fools its enemies. To find caterpillars, many predators home in on the smell of the caterpillar's feces. But skipper caterpillars do something quite sneaky—they shoot fecal bullets (pellets of its solid body wastes) up to six and a half feet away from their bodies. As a result, when a predator zeroes in on the pellets, it won't find a caterpillar.

the use of a simple tool, such as a clam rake, will help you identify their location. By dragging a clam rake behind you as you walk through the shallow water, you can identify places where clams are hiding. The tines of the rake will scrape over partially buried clams revealing their location. You can then use the rake or reach down into the water to pull a clam out of the sandy bottom.

Accomplished clammers will also use another, admittedly ancient, method of identifying where clams are hiding. Sometimes known as treading, clammers wade through shallow water—particularly in areas where the bottom is somewhat soft or muddy. Using their feet as sensors, the clammers detect partially buried clams simply by feel. Many clammers resort to using water shoes, since they are fairly light and very flexible. As you might imagine, dragging your tootsies through the muddy bottom of a bay might reveal all sorts of critters and other marine life that you might not want to reveal . . . or feel . . . or even deal with. It also subjects the clammer to the dangers of scraping one's feet over broken shells, rocks, and other abrasive (and really dirty) objects that could cause some serious harm to one's feet and, ultimately, ruin an entire day at the beach.

Tools of the Trade

One of the great things about clamming is that it is a relatively inexpensive way to enjoy yourself, commune with nature, and get a dinner in the bargain. In short, clamming is a very cheap form of entertainment. Your clothes can be the rattiest and most well-worn items you can drag from the back of your closet. An old sweatshirt or T-shirt, a pair of old pants or a swimsuit, and a comfortable pair of worn-out, slightly disintegrating and ancient sneakers will complete your clamming clothing ensemble.* This is not the time to make a fashion statement; nobody will be looking at you (they're all looking for clams), and the more comfortable you are (take note of the daytime temperature), the more enjoyable your clamming experience. Bottom line—make sure your shoes are old and comfortable, as you'll be getting them wet and dirty.

While it's always possible to dig clams out of the sand with your hands, you will discover that after a while your hands will become sore and the clams a little more tenacious than you originally thought. And, if you are approaching the same age as the author of a certain clam book, you'll also discover that "hand farming" does things to your back that will guarantee

---

* Right now, dozens of New York fashion designers are cringing at these very words.

195

frequent visits to your local chiropractor, massage therapist, or orthopedic surgeon.

> **FAST FACT:** Back pain is the second most common neurologi-
> cal ailment in the United States (only headache is more common).
> Americans spend more than fifty billion dollars each year on remedies
> for low back pain.

For those reasons, it's always a good idea to have one or more clam-ming tools at your disposal. These can be obtained at a local bait shop, sporting goods store, or marine supply company. Clamming tools can also be ordered online (they are available through various retailers, including Amazon.com). I've often found that by scrounging through the shelves in the back of my garage, I can come up with some appropriate tools (such as a garden rake, trowel, or hand rake) that will work in a pinch when I've forgotten to order the right tools online. Be aware that it is quite easy to spend a fortune on a set of very expensive clamming tools, thus mitigating the effects of an inexpensive clam chowder or clambake on the beach. My rule—always go for cheap; the meal you make with your harvest will taste all the better.

Nevertheless, there is one tool every clammer should have in his or her pocket (not literally, of course). That would be a clam rake. A clam rake is similar in shape and appearance to a garden rake, but there are some significant differences. For example, a clam rake will be lightweight, with tines that are two to four inches long. Clam rake tines are longer and thinner than those on a garden rake. As a result, they allow a clammer to slice through the sand, rather than trying to move the sand (as would be the case with a garden rake). In turn, there is considerably more leverage with a clam rake. Ultimately, the clam rake helps to identify the location of clams, especially when the tines scrape over any subsurface clamshells.*

---

* As I write this chapter, I note that clam rakes are priced at anywhere between twenty-five and sixty dollars. Perhaps an old garden rake might be a suitable alternative after all.

Close-up of a commercial clam rake.

FAST FACT: Most clams, depending on the species, are buried just under the surface of the sand. However, there are some clams—most notably the geoduck—that can burrow in the sand up to four feet deep.

Another useful tool is a clam pick, which is simply a smaller version of a clam rake. A clam pick will have either two or three tines and a relatively short handle. A suitable alternative from your garden shed would be a hand rake. Be aware, however, that as with the case in clam rakes versus garden rakes, a hand rake has tines that are somewhat thicker than those on a clam pick. Nevertheless, if you are planning on spending some time on your hands and knees searching for steamers, a clam pick or hand rake would be an excellent tool to have.

Another tool used by many clammers is a hand trowel. Since you may be spending some time on all fours, a hand trowel is an excellent tool to dig for clams. I prefer plastic trowels simply because there's less danger in breaking the shells of any clams you unearth from the sand. Also, a plastic trowel is cheaper to purchase. However, from time to time, especially when clamming in rocky areas, you may discover that a stainless steel trowel is more suited to the task.

If you'd like to prevent extended time on your hands and knees, you may want to invest in a clam shovel. While the retail price of these instruments can be considerable—I recently saw one priced at about seventy-five dollars—they can be invaluable in digging for deep-buried clams. I've discovered, however, that a regular garden shovel (with a rounded tip, rather than a pointed one) works as a very suitable alternative. You don't need a shovel with a large head; turning over all that sand will, in the long run, tire you sooner than you might imagine. A small and narrow head is much easier to sink into the sand or mud. Just make sure the handle is sufficiently long to provide the appropriate leverage.

The final piece of equipment you will need is a container in which to keep your clams. There are a wide variety of options here—everything from an old plastic bucket to an elaborate wire mesh basket. Look through any catalog and you'll discover an infinite range of possibilities (and an equally infinite range of prices). I tend to prefer a wire mesh basket. It's lightweight, easy to handle and transport, and holds a considerable number of clams. Wire baskets have the added advantage of not rusting or corroding. And they can also be easily dipped into the water to wash off any sand you may have captured along with the clams.

Some folks prefer cloth mesh bags to keep their clams. These can be tied to your pants or hung over the side of a boat or pier to allow the clams to spit out any sand or grit in the water. They can be obtained from almost any local dive shop. Some folks just use an old burlap bag to hold their clams. These are easy to find in most hardware stores and are considerably cheaper than some of the clam-holding contraptions currently on the market (with prices up to fifty dollars and more).

If you're planning to hunt for geoducks or razor clams, a clam tube is almost a "must." Clam tubes can be understood by a simple scientific demonstration, one you may have done as a kid. Take a drinking straw and dip it into a glass of water. Cover the top of the straw with your finger and lift the straw out of the water. Look closely and you'll note that there is a column of water trapped inside the straw. Take your finger off the top of the straw and the water flows out of the straw. Why? Air pressure! When your finger is over the straw, there's no air pressure inside the straw and the liquid remains. Take your finger off, and air pressure will cause the liquid to quickly exit the straw.

Clam tubes work in exactly the same way—but only on very sandy beaches. A clammer pushes the clam tube down into the sand, plugs the hole, and then pulls up. The sand and water in the tube (along with any bivalves) comes up. The clammer simply unplugs the tube and everything inside comes out.

Using a clam tube to dig razor clams.

At this point you may be wondering if clamming is a very expensive hobby (with lots of expensive tools), or whether it is something you could get into with a minimum expenditure of the family fortune. I am always mindful of the classic maxim—K.I.S.S. (keep it simple, stupid). Thus, I would suggest only two items for your next clamming venture: a clam rake and a (cheap) container in which to place your tiny treasures. You can certainly add some other tools to your retinue, but please keep in mind their costs and the fact that you are going to have to lug them around wherever you decide to forage for clams. Fewer dollars and less weight are the critical criteria I employ.

Clamming Regulations

It would be nice if you, like your prehistoric ancestors, could walk out on any beach you wanted and begin digging for clams for the family supper. But, alas, that is not the case. As you might expect in any pleasurable endeavor, there are always governmental officials hanging around with bulging clip-boards stuffed with multiple sheets of mind-warping rules, mind-numbing regulations, and mind-blasting laws. Albeit, those rules, regulations, and laws have been enacted to prevent the wholesale destruction of clam beds in a particular area or (heavens!) as a way to deter the infiltration of out-of-state residents who don't respect the local environment and would like nothing better than to pilfer all the good seafood from the locals.

Since it is always a good idea to stay on the right side of the law, you should always check with the necessary officials to ensure that you will be clamming in the right place at the right time and not taking more than your permitted allotment. Suffice it to say, different states, different towns, and different communities have varying regulations on the books to ensure that everyone is in compliance.

FAST FACT: In most states, anyone fifteen years of age and older will need a clamming or shellfish license.

First, check with the local authorities: a local town hall, county agency, or other community entity in the specific area you wish to clam. I've always found it advantageous to stop in at a local bait shop or fishing tackle store and inquire about any local rules and where to obtain the necessary per-mits (often available right in the store). Ask the locals about permit or licensing regulations, quota and legal-size rules, and any other guidelines or restrictions. Always make sure you are in compliance with any and all rules, and you can be assured of a worry-free and most enjoyable clamming adventure.

Clam Conservation

"I'm an accidental environmentalist. After I retired, I wanted to give back to my community and so I got involved in helping to establish ReClam the Bay right here on Barnegat Bay." says Rick Bushnell.

A full head of silver gray hair, a trim physique, a warm sense of humor, and an ever-present smile identify Rick as the next-door neighbor you always wished you had. While I wasn't about to ask him to loan me his lawnmower or help me clear out the brush behind my house, he was more than eager to share his expertise on shellfish farming in general and the impact of clams (and clam education) on the tidy ecological niche known as Barnegat Bay.

It's a brilliant spring day as Rick and I bounce around Surf City, Beach Haven, and Ship Bottom, New Jersey, in his well-traveled pick-up truck. Today, I've driven about 200 miles eastward from my home in Pennsylvania to the Jersey shore to talk with Rick about ReClam the Bay, an active and enthusiastic environmental organization over which he presides. ReClam the Bay is a group of dedicated volunteers focused on improving the health and resiliency of Barnegat Bay through the growing of about one million shellfish each year along with numerous educational outreach programs.

Rick Bushnell at an upweller.

201

"I learned to sail on Barnegat Bay," Rick says. "I'm a licensed captain and I just love to eat clams. After retirement, I took some classes on shellfish gardening from Rutgers University."

"Afterwards," Rick adds with palpable pride, "a group of us—retired businesspeople, attorneys, teachers and such—decided to form an educational outreach group that would contribute to the sustainability of the Bay through various involvement programs."

FAST FACT: Barnegat Bay is part of the Atlantic Intercoastal Waterway. Situated in Ocean County, New Jersey, and fronted by Long Beach Island, it is approximately thirty miles long. It was originally charted by the Dutch in 1614, who named it "Barendegat," or "Inlet of the Breakers." In modern times, the many seaside towns lining the bay (from Bay Head in the north to Beach Haven in the south) have been popular tourist destinations for generations of families from throughout the eastern seaboard.

"At one time —the forties, fifties, and sixties—there were nine hundred baymen* making a living on the bay," Rick says. "However, as the clam population dropped, those jobs began to disappear. Now, there may be only fifteen or twenty baymen making a living from the bay. The fifteen or twenty that are left are [harvesting] clams through aquaculture; but they're also supplementing their stocks a little with wild clams. However, there's no reason why this bay can't be an enormous economic success."

Rick points out that 97 percent of the water is "open for harvest," meaning the shellfish are safe to eat because the water meets or exceeds stringent federal and state standards. As he tells me, the problem now is that the shellfish population is not self-sustaining for a variety of reasons. One of the chief reasons is that there is too much nitrogen going into the bay. The excess nitrogen causes algae blooms that smother the habitat the shellfish need to survive.

Rick echoes a theme persistent among long-time residents who have made the bay their home and, in many instances, their livelihood: Barnegat Bay is in trouble. A serene and picturesque tourist destination for tens of

---

* First used in 1641, the term "bayman" or "baymen" refers to fishermen (and fisherwomen) who make their living primarily in or on a bay. Around Barnegat Bay, the term refers primarily to those individuals who made their living from clamming or shellfish harvesting.

thousands of visitors every year, Barnegat Bay, like many other seaside eco-systems up and down the East Coast, has become a victim of its own success. Tracts of houses spread out in all directions and looming shopping centers and expansive malls line the multi-laned highways that funnel traffic into the shoreline communities. Over the years, multiple housing projects and a plethora of businesses have encroached on more and more land—eroding the natural habitat and severely altering the landscape. The once-pristine waters of the bay have been overloaded with increasing levels of nitrogen—the aftereffects of overfertilization, commercial pollution, increased traffic, and rampant development. Consequently, fish and shellfish populations have been severely reduced through a simple but deadly equation: more stuff is being dumped into the waters than can be naturally filtered from those waters. As a result, biodiversity is disrupted, plants and animal popu-lations are severely affected, and livelihoods are forever altered.

Barnegat Bay is in trouble!

In his book, *The Bayman: A Life on Barnegat Bay*, author (and former clammer) Merce Ridgway writes the following:

> The flow of fresh water from mainland streams, along with the tidal flats (which over time became marshes) created an ideal habitat for many different kinds of birds, animals, fish, and shellfish. It was a great natural hatchery, undisturbed by man until about ten to fif-teen thousand years ago when the native [sic] Americans began to harvest and enjoy the waters and land that made up the ecosystem of Barnegat Bay.
>
> The original people were good keepers. The early settlers found the bay to be bountiful. From the early 1900s, when coastal develop-ment really began, until modern times, there has been a great decline in the quality of life found in, on and around Barnegat Bay. By the late 1950s, the environmental decline of Barnegat bay had become dramatic, and easily recognized by seasoned baymen of the area. If we were to use a graph, we would see a line that is fairly steady for thousands of years. In the last fifty to eighty years it drops nearly off the chart. If it were the stock market, there would be a panic!*

* Merce Ridgway, *The Bayman: A Life on Barnegat Bay* (Harvey Cedars, NJ: Down the Shore Publishing, 2000).

Rick is unequivocal in his assessment: "The biggest problem in this bay is too much nitrogen—eutrophication (overstimulated algae growth) caused by fertilizer and other runoff." Rick tells me that there's a lot of atmospheric deposition (from distant factories) as well as car and truck exhaust. What happens, then, is that the microalgae spikes because of the overload of nutrition. And then the microalgae crowds out lots of things, not the least of which is eelgrass. Eelgrass is important because it's where native baby clams, and many other bay creatures, live. As a result, there is a direct relationship between the reduction in the clam population and the reduction in eelgrass. While there can be a whole lot of reasons around that, much of the reduction in the eelgrass generally comes from reduction in sunlight to the eelgrass, and that in turn generally comes from overnitrification. And, when microalgae dies and sinks to the bottom, it smothers whatever is underneath it.

Rick guides his truck into a decommissioned Coast Guard station in Beach Haven where ReClam the Bay stores two of their three boats. He has brought me here to profile one of the nine upwellers the group uses to grow clams.

An upweller, in simple terms, is a systematic collection of approximately a dozen or more five-gallon buckets—each with screening on the bottom (instead of plastic). Each bucket has a hole on its side near the top. A group of the buckets is placed in a large tank, and there are two rows of about six to eight buckets on either side of a long trough. The hole in the side of each bucket is connected to the trough with a short plastic pipe. Bay water is pumped into the bottom of the tank, where it flows upward through the mesh bottoms of the buckets (the reason why it's called an upweller) and over a mass of baby clams in each of the buckets, eventually exiting through the side hole on the top. As you might imagine, the constant flow of water provides the clams with a ready supply of food and other nutrients. Just as important, this bubbly mechanical ecosystem offers the public hands-on experiences in clam aquaculture. (Watching wide-eyed kids pull up and examine a handful of baby clams the size of M&Ms® is a visual experience like no other.)

As we walk around the upweller, Rick tells me that ReClam the Bay normally purchases about one million clam seeds a year (at a cost of about four thousand to five thousand dollars), typically in late spring. Those tiny seeds (each one is about two millimeters long—approximately the same size as a single poppy seed) are loaded into all the upweller buckets (in various Barnegat Bay shore towns), where they are constantly bathed in a flow of bay water. There, they spend the summer consuming an unlimited supply of phytoplankton and growing out to three times their original size. Readers with kids will note that this would be akin to depositing your teenage son in front of the local steak house or burger joint and telling him, "Eat all you want. I'll be back for you in four months."

Sometime in the fall, the tiny clams are taken out into the bay and planted on small underwater plots (approximately fourteen by twenty feet) leased from the state. Each plot is covered by quarter-inch flexible plastic screening to protect the vulnerable clams from numerous predators (e.g., skates, rays, crabs) lurking in the bay. The young clams remain in these plots until they reach market size, which is anywhere from two to four years later. Rick notes, with a certain amount of parental pride, that the mortality rate for ReClam's clams is less than 1 percent.

"Adult clams do very well in this bay," Rick says with proverbial buttons bursting off his shirt. "When we grow them in the upwellers, we get a very high yield [and] a very low mortality. Then when we put them in the bay and come back to check their growth—it's usually pretty good."

A clam upweller.

Baby clams from an upweller.

But Rick is also quick to point out that the mission of ReClam the Bay is not solely to repopulate Barnegat Bay with new stocks of shellfish, even though those populations provide significant environmental and economic benefits to the region. As Rick puts it, the annual culturing of over one million clams is simply a means to an end, as evidenced in the organization's mission statement.

ReClam the Bay's Mission Statement

ReClam the Bay* is an organization dedicated to providing education and awareness about the environmental benefit of shellfish filtering, feeding, and cleaning our estuary. Our mission is to involve the general public so they will understand that the quality of the water in our estuary, and the quality of the shellfish we eat, are really their responsibility. By involving the public in the care, feeding, and life cycle of these fragile creatures we believe that our citizens will better understand how working with the shellfish can help to clean up the environment and keep it clean.

—ReClam the Bay website (www.relclamthebay.org)

---

* The inspiration for ReClam the Bay came from Rutgers Cooperative Extension of Ocean County. According to them, "The best way to reclaim the bay is to reclam the bay."

"What [Hurricane] Sandy showed everybody is that we're all in this together," Rich says with both intensity and a smile. "At some point you realize that, just as we've been displacing the ducks and the foxes that used to live on these islands along with the clams and fish in the bay, we saw that Mother Nature can displace us. So, we're all in this together; we're sharing this environment."

A conversation with Rick is both informative and comfortable. As two weathered troopers in our late sixties, he and I discover many commonalities in our lives: growing up in a coastal community, developing a love and appreciation (as well as an appetite) for various marine creatures, and spending long summer days out on the water in various types of watercraft. The major difference between us is that he used to cruise the Jersey shore as a teenager, while I was cruising the California coast. Nevertheless, a proclivity to educate others and a passion for the sea and its denizens bind us in a special camaraderie.

As part of its outreach efforts, ReClam the Bay informs the public about both the intrinsic and extrinsic value shellfish, specifically clams, can add, not just to Barnegat Bay, but to any seashore environment. ReClam's numerous educational projects and programs help to inform the public about the significant contributions made by clams—how they help stabilize the seafloor and shoreline by absorbing wave energy, how they filter the bay's water, how they add to the biodiversity of a marine ecosystem, and how they provide food (and in many cases, jobs) for those who live near, or travel to, seaside communities.*

After several stops to check on other upwellers and point out various sites on "The Clam Trail," Rick directs the truck back to his seaside house. As we sit in the kitchen (with an absolutely stunning, 130-degree view of Barnegat Bay), I ask him what he and ReClam the Bay would like both tourists and residents to learn after doing any of the following activities sponsored by the organization: visiting one of the group's upwellers (to grow and care for clams); attending a lecture ("A Clam is a Man's Best Friend"); helping kids collect "clam cards" with fun facts that teach the heritage and ecology of Barnegat Bay; taking an eco-kayak tour on the bay in the summer; dragging a net across the bay bottom to find all the different creatures that live in the bay; or trekking through various towns searching for oversized, multi-colored, hand-painted clams embossed with fascinating facts.

---

* ReClam the Bay began in 2005 with about forty volunteers. Today they have more than 70 active members with an additional 450 on their mailing list. On average, they speak to, or influence, more than 10,000 people each year.

"We want people to get a general understanding of water quality and an understanding that you can do things to improve water quality. We want to help people to feel a connection to living things. And, of course, if they're growing their own clams, then that connection is made. As a result, they may alter their behavior."

That Rick and ReClam the Bay are passionate about their mission is a clear understatement. As I look outside, I am met by what once was a naturally sculpted ecosystem teeming with life and professed sustainability. Now that system is being challenged by massive human forces that have incalculable impact and enormous power. It seems almost incomprehensible that a tiny critter—a once-ancient, calcium-encased invertebrate without a brain—can provide the key to the future of this environment. Or to use Rick's words: "What we're doing is an example of a way for people to get intimately involved in living creatures that depend on our environment—what we do to the environment—and can enhance the environment if we work with them. So, the resounding message is—be outside and understand nature and how it's all connected and how you can make things better . . . and nature can make things better for you. It isn't just all about the clams, it's about understanding the role of that little organism in the big scheme of things."

One of the brochures ReClam the Bay distributes to the public paraphrases that essential idea quite well, "They support us, so we ask you to support them!"

Chapter 12

They're Good for You!

One cannot think well, love well, sleep well, if one has not dined well."
—Virginia Woolf

S AY "FOURTH OF JULY," AND MOST PEOPLE WILL think of sparkling firework displays, long parades with marching bands, family picnics in the park, droning speeches by droning politicians, and loads of mid-summer fun! Other people will think of good old American hot dogs—or, more specifically, the famous annual hot dog eating contest sponsored each July 4th by Nathan's Hot Dogs on Coney Island, New York. For this popular event (now televised live on ESPN), contestants try to outdo each other (actually, out-eat each other) to see who can consume the most hot dogs (with buns) in ten minutes. The current world record (as of this writing) is sixty-nine hot dogs consumed in ten minutes. It is held by Joey Chestnut of San Jose, California, who has won this contest (as well as other major eating contests) several times.

> **FAST FACT:** Chestnut began his competitive eating career in 2005 with a record performance in the deep-fried asparagus eating championship. He consumed 6.3 pounds of asparagus in 11.5 minutes (that's without hollandaise sauce).

Lest you think rapid and prodigious food consumption is an art form practiced exclusively by starving teenage boys or fraternity pledges, please be advised that there is serious competition, administered and regulated by the Major League Eating (MLE) organization, in any number of professional competitive eating events around the country. Consider the following world records. (Please do not try these at home!)

FOOD	AMOUNT	TIME	RECORD HOLDER
Baked beans	6 lb.	1 min., 48 sec.	Don Lerman
Blueberry pie	9.17 lb.	8 min.	Patrick Bertoletti
Butter	7 sticks (1/4 lb. ea.)	5 min.	Don Lerman
Cow brains	17.7 lb.	15 min.	Takeru Kobayashi
Eggs	141 hard boiled	8 min.	Joey Chestnut
Ice cream	1 gal., 9 oz.	12 min.	Cookie Jarvis
Mayonnaise	128 oz.	8 min.	Oleg Zhornitsky
Pig's feet	2.89 lb.	10 min.	Arturo Rios, Jr.
SPAM	6 lb.	12 min.	Richard LeFevre
Watermelon	13.22 lb.	15 min.	Jim Reeves

Not to be outdone (or outeaten), several clam-eating contests are held annually around the country. While the press has not sufficiently warmed up to the idea of brisk clam-eating competitions to the same extent as they have to the swift ingestion of, say, a combination of meat trimmings, fat, salt, garlic, and nitrates (a.k.a. hot dogs), let it not be said that our favorite bivalves haven't established themselves as worthy of victual contests. Case in point: below are two recent winners (and potential record holders) for conspicuous clam consumption.

Male: At the 19th Annual Highlands Clam Festival held in Highlands, New Jersey, on August 4, 2013, clam champion Ed Woodruff ate ninety-two clams in five minutes (that's one clam every 3.3 seconds). Shortly after the competition, Woodruff stated that he "felt full and very salty." (Apparently, Ed is also a champion of understatement.)

Female: On May 31, 2010, at the World Cherrystone Clam Eating Championship on Long Island, New York, winner Sonya Thomas (a.k.a. "The Black Widow") scarfed down 312 cherrystone clams in six minutes. She also walked away with the top prize of $1,250—that's a lot of "clams"!

As you can see, the ingestion of clams is not without honor and celebration. But let's not dwell on the tactics of winning the top prize at a local summertime contest for consuming massive quantities of bivalves. Rather, let's take a look at the nutritional benefits of these underwater delicacies—for our ancestors as well as for ourselves.

Clams as Brain Food

As we've discovered earlier in this book, our forefathers were as enamored of clams as we are today. That massive quantities of clams have been eaten by ancient peoples throughout history is a given. Clams have proven themselves to be a significant dietary staple, particularly for those folks who had the great fortune to live near enormous bodies of water—the ocean, for example.

Indeed, clams are a fact of life for many shoreline inhabitants. What is not so clear is how those clams may have influenced human evolution—specifically, the growth and development of hominid brains. For many anthropologists that's a major conundrum. You see, in the field of anthropology there is a significant and ongoing debate about how human brains developed over the millennia. In short, the evolutionary growth of our "gray matter" is a persistent puzzle among many scientists. Unanswered questions include the following: What spurred the expansion of our brains over time? What influenced their growth? How did our brains get from somewhat small (our ancestors . . . and most politicians) to somewhat large (you and your relatives)?

FAST FACT: Two million years ago our earliest human relative, *Homo habilis*, had a brain roughly half the size of our brains today.

One intriguing theory comes from Dr. Stephen Cunnane, a metabolic physiologist at the University of Sherbrooke in Sherbrooke, Quebec, Canada. According to Cunnane, the brains of modern humans are, most likely, the result of a shore-based diet that provided essential nutrients necessary to its development.

"Anthropologists and evolutionary biologists usually point to things like the rise of language and tool making to explain the massive expansion of early hominid brains," says Cunnane. "But this is a catch-22. Something had to start the process of brain expansion, and I think it was early humans eating clams . . . from shoreline environments. This is what created the necessary physiological conditions for explosive brain growth."

Cunnane argues that healthy human brain development, and its evolution, is critically dependent on certain metabolic factors. Consider, for example, the energy needs of a modern human brain. A developing newborn's brain consumes an incredible 75 percent of that infant's energy needs each day. To feed that energy demand, babies tap into a built-in reserve of energy—their baby fat. A newborn's baby fat accounts for approximately 14 percent of its birth weight, an amount similar to that of its brain.

Cunnane makes the case that baby fat is directly related to hominids' evolutionary brain expansion. He contends that fat is the direct result of babies' mothers "dining on shoreline delicacies like clams and catfish."

"The shores gave us food security and higher nutrient density," says Cunnane. "My hypothesis is that to permit the brain to start to increase in size, the fittest early humans were those with the fattest infants."

As proof, Cunnane points to ancient shoreline environments that offered easily accessible and readily available food supplies. Some of the most sustainable and longest-lasting human communities were established in these environments . . . and have remained so to this day. It was these shorelines that provided essential brain-boosting nutrients and minerals that launched *Homo sapiens* brains past their primate peers, concludes Cunnane.

So, clams and increased brain size. Hmmm, I think I'll have some more chowder, if you please!

How Healthy Are Clams?

Remember when you were a kid and your mother or father used to say to you at the dinner table, "Eat your (blank), they're good for you!" or "If you don't eat your (blank), you'll never be able to (blankety-blank)." It didn't make any difference what the food was—brussels sprouts, peas, carrots, broccoli, boiled cabbage, liver, or whatever—eating it (at least according to your parents) would ensure that you would be forever healthy, strong, and fit.* Not eating your (blank) would either doom you to a life of nutritional hell or get you sent up to your room.

Well, I'm going to exercise my parental and professorial authority and tell you to eat lots of clams. Why? They actually have many nutritional benefits that put them head and shoulders (assuming they had heads and shoulders) above many other types of foods. They are, in fact, dietary super-stars, offering you all sorts of benefits that your typical steak and potatoes meal could never hope to provide. Let's take a look.

Iron

Growing up, I remember watching Popeye cartoons. When faced with some dire calamity or big brute (Bluto), Popeye would quaff a can of spinach and instantly be endowed with superhuman strength sufficient enough to fight off any and all assailants or menacing terror invading the neighborhood. The not-so-subtle message to impressionable youths watching those Saturday morning shows was that if we made spinach a regular part of our diets, we, too, could beat up any bullies roaming the 'hood or effectively deal with any potential havoc that might befall our hometown.

However, it was not clear why spinach was so good for us (and certain cartoon characters). It wasn't until high school biology that we learned that spinach is packed with iron, and iron is one of those nutritional essentials that every human body needs. Happily, we don't have to consume endless

---

* When I was young I was told that eating bread crusts would make my hair curly. Now, well into my late sixties, I have a considerable "chrome dome." Hmmm, I'm wondering if my parents were just feeding me some false and misleading information.

cans of spinach to get our iron. In fact, our minimum daily adult requirement can be obtained from a mere three ounces of clams.

Let's put it this way—in the iron department, clams shine! In fact, clams are one of the most iron-rich foods you can eat. Three ounces of clams (roughly eight to nine small clams) is equivalent to three ounces of beef or chicken liver. (And, let's face it, wouldn't you rather be eating clams than liver?) Those three ounces of clams contain approximately 3.5 milligrams of iron.

FAST FACT: Here are the Recommended Daily Allowances of iron for various ages:

Children	1–3 years	7 mg
	4–8 years	10 mg
	9–13 years	8 mg
Females	14–18 years	15 mg
	19–50 years	18 mg
	51+ years	8 mg
Males	14–18 years	11 mg
	19+ years	8 mg

Why did the iron in spinach give Popeye so much strength, and why is it important in your daily diet? For one very important reason—without sufficient iron, your body can't produce enough hemoglobin, a substance in red blood cells that makes it possible for them to carry oxygen to the body's tissues. Without sufficient hemoglobin, you may feel weak, tired, and irritable. In addition, iron is necessary for preventing iron-deficiency anemia. According to the folks at the US Department of Health and Human Services, most women and young children in the United States are at risk for inadequate iron intake. So, bottom line, eat clams and get your iron!

Protein

You don't have to be an Olympic athlete to know that protein is an important and essential ingredient in any healthy diet, and lean proteins are even better. You could eat lots of lean meat or chicken to get your daily requirement of protein, but why not consider clams as a suitable source for your protein needs? You see, clams have more protein content than other forms of seafood (such as those oyster creatures). In addition, they also have considerably more vitamins and minerals than other protein-rich foods (such as those chicken creatures).

> **FAST FACT:** Looking for lean proteins (besides clams)? Here are some other possibilities: canned salmon, eggs, ground sirloin, lentils, low-fat cheese, pork loin, soybeans, sushi, turkey breast, and yogurt.

Let's look at it this way: a three-ounce serving of steamed clams (eight to nine small clams) provides your body with twenty-two grams of protein (equivalent to 44 percent of the daily value based on a 2,000-calorie diet). In addition, those eight to nine clams have only 126 calories and less than two grams of total fat. The big advantage of clams (as it is with other high-protein, low-calorie foods) is that they can help you lose weight or prevent significant weight gain. That's because lots of protein helps you feel full longer after a meal. As a result, you tend to eat fewer calories at the next meal. Yeah!

Good Fat

Pick up any newspaper or popular magazine and somewhere you're likely to find the word "fat" in it. Actually, you'll discover many instances of fat in almost any periodical in the supermarket checkout lane. Fat, fat, fat—it seems to be the word of choice and certainly a word that the producers of some reality shows like to throw at us with increasing regularity. It's gotten so bad that the word "fat" has been deemed evil, horrible, terrible, awful, and downright bad. But, let's give the word a break; it needs it! Let's talk about good fat.

You see, clams are a major source of omega-3 fatty acids. In fact, a three-ounce portion of clams contains 117 milligrams of eicosapentaenoic acid (EPA) and 174 milligrams of docosahexaenoic acid (DHA)*. EPA and DHA are often referred to as the "good fats."† The good fats regulate your

---

* Not to worry—I had a hard time pronouncing these acid names, too!
† Instead of "good fats," I like to think of these as "damn good fats."

hormones, reduce inflammation, increase your brain power (see above), and may reduce your risk for heart disease by lowering your blood triglyceride levels. Unfortunately, most people do not get enough good fats in their daily diet. A serving of clams can go a long way to ensure that you get the (good) fats you need.

An additional benefit of eating clams to get your daily requirement of omega-3 fatty acids is that clams, generally, are free from mercury, unlike several other kinds of seafood. As you may know, mercury can have serious deleterious effects on human beings—particularly unborn children. That is why pregnant women need to be cautious about the seafood they eat.

Cholesterol

Cholesterol, for all the bad press it has received, is necessary for proper functioning of your body. Cholesterol is a fat-like substance produced by your liver and also found in certain foods, such as animal meat and selected dairy products. Your cell walls need cholesterol in order to produce hormones, vitamin D, and the bile acids that help to digest fat. But too much of a good thing can be a bad thing. In short, a little cholesterol is good, but too much cholesterol can be horrible for you.

If you or I eat too many fatty foods, plaque tends to build up in our arteries. That plaque constricts the arteries, narrowing the space for blood to flow to the heart. In time, this may lead to atherosclerosis (hardening of the arteries), which can lead to heart disease and other types of cardiac difficulties.

But it's important to know that not all cholesterol is bad. Just like with "fat," there is both "good" cholesterol and "bad" cholesterol. "Good" cholesterol (also known as high-density lipoprotein, or HDL) helps the body get rid of bad cholesterol in the blood. The higher the level of HDL cholesterol, the better. "Bad" cholesterol (low-density lipoprotein, also known as LDL), on the other hand, can cause buildup of plaque on the walls of arteries. The more LDL there is in the blood, the greater the risk of heart disease.

FAST FACT: The American Heart Association recommends that you limit your average daily cholesterol intake to less than three hundred milligrams. If you have heart disease, you should limit your daily intake to less than two hundred milligrams.

So where do clams fit into all this talk about cholesterol? Well, a three-ounce portion of raw clams contains a mere 29 milligrams of cholesterol, or 10 percent of the maximum amount that healthy adults should have in a day. Now, if you were to eat three pounds of clams (as opposed to three ounces), you would be getting 319 milligrams of cholesterol—well above the maximum amount any sane adult should be ingesting. Let's just say, as far as clams and cholesterol are concerned, moderation might be the most appropriate course.

Calories

With only two grams of fat and fewer than one hundred calories per three ounces, eating clams is a great way to diet while providing for many of your nutritional needs. If you're counting calories (or, more specifically, counting clams), here are some figures to consider:

- 1 small clam = 7 calories (12 percent fat, 15 percent carbs, 73 percent protein)
- 1 medium clam = 11 calories
- 1 large clam = 15 calories
- 3 ounces of raw clams = 63 calories
- 1 cup with clams and liquid = 168 calories

FAST FACT: To burn off sixty-three calories (three ounces of clams), you would need to do one of the following: walk for seventeen minutes, bike for ten minutes, jog for seven minutes, or swim for five minutes.

Vitamins and Minerals

If you're looking for a vitamin and mineral powerhouse, you can't go wrong with clams. They're vitamin and mineral factories of the highest order. Eat some clams every day, and you'll be fueling your body with some of the "big hitters" as far as vitamins and minerals are concerned. Let's take a look.

Clams are *very high* in the following:

- **Vitamin B12** is important to many bodily functions. People who don't get sufficient Vitamin B12 in their daily diet may experience one or more medical symptoms, including tiredness, weakness, constipation, loss of appetite, weight loss, megaloblastic anemia, nerve problems, difficulties with balance, depression, confusion, dementia, poor memory, and soreness of the mouth or tongue. To prevent those things from happening to our bodies, we need about 2.4 micrograms of vitamin B12 each day. Vitamin B12 also helps insulate your brain cells as you age (this is particularly helpful information for senior citizen authors).

- **Vitamin C** is an antioxidant that protects your body cells against damage, helps wounds to heal, fights infections, promotes healthy bones, teeth, gums, and blood vessels, and aids in the absorption of iron. While many nutritionists advocate lots of citrus fruits (packed with Vitamin C) in your diet, you might also want to consider adding some *Mercenaria mercenaria* as well.

- **Manganese** is an essential nutrient for your body. It is involved in many chemical processes, including processing of cholesterol, carbohydrates, and protein. Manganese is also used in the treatment of osteoporosis, anemia, and symptoms of premenstrual syndrome.

- **Phosphorus** is necessary for growth of bones and teeth. It assists in muscle contraction, kidney function, heartbeat regulation, and nerve conduction. Children need about five hundred milligrams a day; adults about seven hundred milligrams.

- **Selenium** is an important trace mineral that helps protect your cells from damage by free radicals. It also helps regulate your thyroid and is necessary for your immune system to function properly.

FAST FACT: "Limey" is a colloquialism frequently used for sailors in the British Royal Navy since the mid-1800s. It is believed to have originated as "lime-juicer," and was later shortened to "limey." It came about as a result of adding lemon or lime juice (good sources of Vitamin C) to the sailors' daily ration of rum to prevent scurvy, a debilitating disease prevalent among sailors who did not have access to perishable fruits.

Clams are *high* in the following:

- **Niacin** is essential for energy metabolism in your cells, the proper functioning of your gastrointestinal and nervous systems, healthy skin, and the release of energy from carbohydrates, fats, and protein. Adults should have approximately fourteen to sixteen milligrams of niacin per day.
- **Potassium** helps maintains your heartbeat and is important in many of your body's metabolic reactions. It balances fluid inside and outside your body cells, thus helping to maintain normal cell function.
- **Zinc** is a mineral necessary for growth, especially during pregnancy and childhood, and for tissue building and repair. Your body needs sufficient zinc to help wounds heal, maintain a healthy immune system, and aid cell reproduction.
- **Vitamin B2**, also known as riboflavin, helps you get energy from carbohydrates. It is important for growth and red blood cell production. As an adult, you should get about 1.7 mg per day.

> **FAST FACT:** You can always tell when you have sufficient vitamin B2 in your diet. Your urine turns a distinctive and brilliant yellow color.

Think about this: Clams are packed with all sorts of vitamins and minerals that will improve your health, extend your life, and promote your overall wellness. The old saying used to be "an apple a day keeps the doctor away." How about, instead, "eat some clams every day and keep *all* the doctors away."

Salt

You were probably waiting for the other shoe to drop. That is, there has to be something bad about clams, at least from a nutritional point of view. Well, I suppose there is. Steamed clams, for example, are naturally salty (big surprise), with 1,022 milligrams of sodium in a three-ounce serving. Those governmental experts hanging around the Centers for Disease Control tell us that healthy adults should limit their sodium intake to about 2,300 milligrams per day. Eat seven ounces of clams, and you've just exceeded your daily limit of salt. And, as you know, an excess of salt can significantly raise your blood pressure—not always a good thing.

Now, repeat after me: moderation, moderation, moderation.

When all is said and done, you'll find that clams are one of the healthiest foods you can eat. Cooked in a variety of ways—steamed, baked, fried, sautéed—they can certainly add to your overall nutritional needs as well as culinary predilections.

Now, I'll put all that nutritional information into an (edible) context. We'll take a standard clam recipe and examine all its nutritional benefits. Let's use one of my favorites:

Clams Italiano

Italian cooking is renowned throughout the world primarily for two distinctive reasons: the quality of the ingredients and the simplicity of preparation (with many recipes limited to just four to eight ingredients). While there are many regional differences (or modifications) both within Italy as well as throughout the world, Italian recipes are easily translated into any culture or any kitchen.

Virtually surrounded by the Mediterranean Sea, Italy is known for fantastic seafood recipes. It should come as no surprise then, that clams are a featured ingredient in many classic recipes. By the same token, wine is often featured in many of the best-known dishes from Italy. Combine clams and wine, and you have the makings of a meal that will delight the whole family.

½ cup butter
5 cloves of garlic, minced
2 cups of dry white wine
1 tablespoon dried oregano
1 tablespoon dried parsley
1 teaspoon crushed red pepper flakes
3 dozen clams, scrubbed clean
rolls or bread for serving

1. Melt the butter in a large skillet over medium heat. Cook the garlic in the melted butter for about two minutes until it wilts. Stir in the wine and season with oregano, parsley, and red pepper flakes.

---

2. Place the scrubbed clams in the wine mixture. Cover, then steam them until all the clams have opened, approximately five to ten minutes. Discard any clams that do not open. Ladle the clams into soup bowls and pour the broth generously over them.

3. Have warm, crusty rolls or a loaf of good Italian bread on hand. Tear off pieces and dip them into the wine clam sauce for an extra treat.

Serves: 2–4

Okay, now that you're done with that delicious repast, let's take a look at what you stuffed into your mouth—nutritionally speaking, of course. Our discussion below, however, makes two assumptions: one, you limited yourself to just one serving and two, that you are on a two-thousand-calorie diet each day. If either of those assumptions is incorrect, please adjust the data below accordingly.

First of all, you consumed about 227 calories, or 11 percent of your daily caloric intake. Approximately 142 of those calories were from fat—an amount that is about 24 percent of your daily intake. Nearly ten grams of saturated fat were in your meal, or roughly 49 percent of your daily requirement. In addition, your single serving had a grand total of 47 milligrams of cholesterol or 16 percent of your minimum daily requirement.

As you recall from our discussion above, clams are very low in carbohydrates; for this particular meal there were only 4.4 grams of carbs, a mere 1 percent of your daily requirements. Your fiber intake was also on the low side, coming in at 0.6 grams or 2 percent of your daily needs. (I'm afraid you'll still need to eat your bran flakes for breakfast tomorrow.) On the other hand, there were 3.2 grams of protein, or 6 percent of your nutritional needs for the day. This was balanced by just 126 milligrams of salt, a measly 5 percent of your daily intake.

FAST FACT: Roman soldiers were sometimes paid in salt. This is where the word "salary" comes from.

Since you're concerned about a well-balanced diet, you might be wondering about the kinds of vitamins in your single serving of "Clams

Italiano." You got 16 percent of your minimum daily requirement of Vitamin A, 7 percent of your daily requirement of Vitamin C, 5 percent of your requirement of calcium, a whopping 37 percent of your iron needs (ladies, take note), 3 percent of thiamin, 9 percent of niacin, a hefty 67 percent of Vitamin B, 5 percent of magnesium, and 4 percent of folate. I think it's safe to say that just one serving of this magnificent dish helps ensure that several of your vitamin needs for the day are being satisfied.

When all is said and done, just look at what you've packed into a single dish: a single serving of "Clams Italiano"! I think we can all agree that clams have the potential for ensuring our health while also satisfying our culinary urges. As our Italian friends would say, *"Bellissimo! Bellissimo!"*

Chapter 13

Culinary Odysseys and Epicurean Journeys

So long as you have food in your mouth, you have solved all questions for the time being.

—Franz Kafka

MY WIFE AND I LOVE TO TRAVEL. WE PARTICU-larly enjoy out-of-the-way places—places without the raucous hustle of tourists or the din of rumbling semis echoing in our ears. We especially enjoy seaside ventures—sandy vistas where we can stroll along the edge of a rising tide, where the sunsets often echo off the pastel colors of a cozy B&B, and where we can collect unusual sea shells and artistic driftwood often strewn across cascading sand dunes populated by noisy gulls and skittering shorebirds.

Those maritime adventures are frequently highlighted by true seafood discoveries, notably in off-the-beaten-track, nonfranchised restaurants where the chef knows better than to serve frozen fish or week-old bivalves. Some of the best eateries are hole-in-the-wall places with rusting signs and museum-quality furnishings. The menus are spotted, the silverware is plastic, the servers genuinely cheerful, and the food unparalled. It is here we enjoy fried clams with just the lightest of batters. It is here we savor steamers that haven't been reposing inside a walk-in cooler for days. And, it is here we rediscover the joyfulness and culinary passion reminiscent of those long-ago family feasts.

Suffice it to say, there are scores of eateries along the east, gulf, and west coasts to satisfy anyone's culinary urges and shellfish predilections. Unfortunately, there are far too many to list in this chapter.

FAST FACT: According to the National Restaurant Association, there were 990,000 restaurants in the United States in 2014. All those restaurants, combined, generated a total revenue of $683.4 billion.

However, if you are on the hunt for a diner, café, or restaurant that serves clams—great clams—then, there are certain protocols you should keep in mind. Sure, you can always check out websites such as yelp.com or tripadvisor.com, but you might also want to keep the criteria below in mind. Admittedly, these are arbitrary standards, and, while they may not be perfect, they have served me well over the years in deciding where I was going to tuck a bib into my shirt, where I was going to dip my *Mercenaria mercenarias* into a Styrofoam container of melted butter, and where I was going to leave my hard-earned money (or "clams").

1. I prefer my clams in casual, funky, down-home places. If more than half the customers are wearing ties or if the wait staff is dressed in anything more elaborate than cheeky T-shirts ("I'm one tough broad, but my clams are oh so tender!"), then you'll probably find me elsewhere.
2. I particularly like places that have been around for more than one generation—family-run joints or mom-and-pop places where everyone in the family is involved and where they've been serving shellfish for decades. If they've been around for half a century and there are still long lines snaking out the door, then that's for me. If they opened last Tuesday and there's only one or two cars in the parking lot, then I keep on moving down the road.
3. Every place must adhere to one cardinal rule: I want places that serve tons of clam dishes every day. That way I know there's a high turnover of clams, meaning lots of clams are being delivered almost on a daily basis. In short, the clams are fresh.
4. I know I'm going to incur the wrath of my friends in the medical profession (as well as certain daytime TV doctors), but a good clam place shouldn't be afraid to use the word "fried." They should not be afraid to post large signs with phrases such as "FRIED CLAMS" or "TASTE OUR DEEP FRIED CLAM CAKES" on various walls of the

establishment. Hey, it's not like you're making "fried" a perpetual part of your nutritional lexicon. You're just treating yourself, right?

5. Any establishment serving bivalves should be within spitting distance (clam spitting distance) of the sea. A restaurant just off the interstate in the middle of the Rocky Mountains and leagues from any ocean is hardly a place to order the freshest of clams. I want to be able to smell the seaweed, observe flocks of raucous gulls, hear some pounding surf, and get a few buckets of sand in my shoes. In short, I want to know the sea is nearby—in close proximity to its denizens, particularly those that now repose on my dinner plate.

6. I steer clear of oversized shopping malls with acres of parking and stores the size of aircraft hangers. If an eating establishment has a Wal-Mart, a chain drugstore, a lumber yard, a European furniture store, a Macy's, and a RadioShack as its neighbors, I typically keep on keeping on (down the road, that is).(See #1, above.)

7. The fact that a restaurant has an old fishing net hanging on the wall (with some shells, plastic fish, and a wooden lobster stuck in it) doesn't automatically make it a great place to order clams. Trust me on this one.

8. One final rule: The largest, brightest, and newest places aren't always the best; and the ones with the fanciest websites may be more expensive than good. Your best bet: Always ask the locals—where do they go for great clams? One of the reasons my wife and I enjoy staying at B&Bs is that the owners always know the great eateries.

Finding great seafood restaurants,especially when you're traveling, can often be a challenge. Check out websites, ask around, see where the long lines are or where the parking lots are crowded. Do your homework, and you will be rewarded with clam dishes that will keep you salivating long after you leave the restaurant.

Clam Shacks

If you're traveling through New England, particularly in the summer months, you'll undoubtedly come upon iconic clam shacks scattered hither and yon up and down the seaboards of Maine, New Hampshire, Rhode Island, Massachusetts, and Connecticut. You may have heard of some of the more famous establishments, such as Woodman's of Essex, The Clam Box, The Clam Shack, Ken's Place, or The Sea Swirl, as well as many others that will satisfy all your culinary urges. These are the places where clams are king—places where clams crowd out everything else on the menu, places where clams are as much a passion as they are an art form, and places that celebrate clams as epicurean royalty (which they are, of course).

In my opinion, to be a legitimate clam shack, an establishment must adhere to certain "standards." As your "official clam guide," I would like to share my bivalve bylaws with any who journey through this gourmet wonderland:

- An official clam shack is usually a stand-alone, weathered, hand-painted, and quite rustic building.
- A certified clam shack must serve a broad variety of clam dishes— everything from clam strips to clam cakes.
- If there is a menu, it is at least several generations old. If there isn't a menu, all the offerings are on hand-lettered signs.
- The lines are long, but the wait is worth it.
- You order your meal from a window or counter. (You are your own wait staff.)
- The people behind the counter are always busy, but always have time for good customer service.
- You go to a second window or counter to pick up your order.
- All the silverware is plastic, and the plates are cardboard.
- Cash only, no credit cards.
- Meals are eaten outside on picnic tables with lots of other people.
- You throw all your "dinner dishes" into a trashcan.
- You always leave with a full belly and a big smile.
- EXTRA—You tweet your best friend: "OMG I can't believe the food I had at _____ Clam Shack. The clams R so ☺!"

IMPORTANT NOTE:

- Not all clam shacks are open year round. Several are seasonal (mid-April to mid-October, for example). Call or check out an establishment's web-site if you're planning to visit at a time other than summer.

> **FAST FACT:** March 31 of each year is National Clams on the Half Shell Day.

Clam Festivals

Clams are always cause for celebration. Clams are all about good times, good friends and, of course, good food. On both the east and west coasts there are festivals and events that will satisfy your culinary impulses just as much as they will satisfy your urges for wild and raucous sand castle-building contests, long and noisy parades, lots of arts and crafts booths, itinerant face painters, stands filled with cheap jewelry, and all manner of food sold from an array of decorative tables and the back end of rusting vans. This is apple-pie, down-home Americana in no-holds-barred festive occasions that last for days and are remembered for years.

> **FAST FACT:** According to Wikipedia, there are 358 regularly scheduled annual festivals (one-day or multi-day) held in the United States each year (almost one festival for each day of the year).

Travel around the country and you'll discover festivals that put clams in their proper place—on a pedestal. These are the times to let loose and embrace clams as you never thought possible. Fried, steamed, broiled, or raw—bushels of clams await your taste buds along endless blocks of display stands, makeshift stages, and bannered stalls. These are the times to embrace clams, cheer clams, toast clams, and high-five clams in all their culinary forms.

Come and be part of the party . . . come and enjoy!

Annual Clam Chowder Cook-Off & Festival

Santa Cruz, California
www.beachboardwalk.com/clamchowder
End of February

Join the fun and excitement as talented chefs from throughout the West Coast compete for the glory of the Best Clam Chowder! More than just a culinary competition, the fun event has featured participants dressed as scuba divers, mermaids, and even as a Clam Fairy.

Annual Essex Clamfest

Essex, Massachusetts
www.capeannvacations.com/festival-event.cfm?id=79
End of October

For more than thirty years this festival has been celebrating clams. Included are more than forty vendors and stands featuring lots of area restaurants where you can sample all the calm delicacies.

The Great Monterey Clam Chowder & Calamari Festival

Monterey, California
www.seemonterey.com/events/food-wine/clam-chowder-calamari-festival
End of May

Two of Monterey County's most loved dishes come together on Memorial Day weekend. Custom House Plaza swarms with comforting cuisines from the sea along with lots of entertainment!

Highlands Annual Clam Festival

Highlands, New Jersey
www.highlandsnj.com/hbp/calendar/html/clamfestb.html
1st Weekend in August

The Clam Festival is a four-day action-packed event featuring the freshest seafood, live entertainment, kiddy rides, games, contests, a beer and wine garden, and more.

Ocean Shores Razor Clam Festival

Ocean Shores, Washington
www.oceanshores.org/clams.html
End of March

It's all about razor clams . . . and you won't leave disappointed from this annual celebration of food and fun!

Pismo Beach Clam Festival

Pismo Beach, California
www.pismoclamfestival.com
3rd Weekend in October

This is the "granddaddy" of clam festivals. You haven't been to a festival, and you haven't tasted clams, until you've come to Pismo Beach. You'll discover more to do here (and eat here) than anywhere else in the country. This is a first-rate clam celebration! It should be on everybody's "bucket list."

Razor Clam Festival

Long Beach, Washington
longbeachrazorclamfestival.com
3rd Weekend in April

Welcome to Long Beach, Washington—home of the Razor Clam Festival.

Bands, pirates, mermaids, games, face painting and, of course, lots of clams to savor make this festival a "must-see" event in the spring.

Yarmouth Clam Festival

Yarmouth, Maine
www.clamfestival.com/home.php
3rd Weekend in July

Yarmouth is like a family reunion on steroids. These folks know how to celebrate clams, and they know how to share their clams with the rest of the world. Put this festival on your "to do" list and you will think you've died and gone to heaven.

Chapter 14

Recipes to Die For

I cook with wine. Sometimes, I even add it to the food.
—W. C. Fields

IT WAS A SCANDAL!

Even by New York standards, it was a heinous event of epic proportions. It was bold! It was daring! It was ruthless! It kept tongues wagging and the presses rolling for weeks, even months, afterwards.

It was the talk of the town!

By all accounts it was a heartless, brutal, and calculating murder—a murder unlike any seen in the city for some time. And the weapon of choice wasn't an ax. It wasn't a dagger. It wasn't a pistol.

The murder weapon was . . . you guessed it—clams. Or, to be more specific, clam chowder (with just a pinch of arsenic for good measure).

Here is how it all unfolded: The year was 1895, the dawn of the Gilded Age. Mrs. Evelina Bliss was a very wealthy widow living in Harlem who happened to have a daughter—one Mary Alice—from her first marriage. Mary Alice, who lived in the nearby Colonial Hotel, could best be described as a "bad seed." You see, she was unmarried, yet she had three children, each fathered by a different man (quite scandalous in the nineteenth century). And, to top things off, there was a fourth child on the way.

Apparently, Mary Alice was desperate for money to feed her growing brood, and she saw a golden opportunity in her mother's vast riches. So, on the evening of August 30, she sent her ten-year-old daughter to Granny's house with a specially prepared bowl of clam chowder. A specially prepared bowl with one extra (and quite fatal) ingredient.

The unfortunate Mrs. Bliss passed on later that night (most likely in a condition that was in direct opposition to her name). When the coroner arrived it quickly became apparent that she had been the victim of arsenic poisoning.

> **FAST FACT:** The symptoms of arsenic poisoning include abdominal pain, diarrhea, vomiting, blood in the urine, cramping muscles, hair loss, dark urine, stomach pain, dehydration, vertigo, delirium, severe convulsions, shock, and ultimately death. Not a pretty picture, you must admit.

Mary Alice was immediately picked up by the police and charged with murdering her mother. All the clues and all the evidence pointed in her direction. The press had a field day; the story was front-page news for weeks. It was clearly evident that the prosecution had a very solid case. The fact that Mary Alice was a single mother was a further indictment, especially in Victorian New York, where single motherhood was earnestly and overtly frowned upon. Even though Mary Alice could trace her heritage to "old money," her wanton ways made her as guilty as sin in the eyes of the public.

But Mary Alice had a very clever lawyer, one who was smart enough to capitalize on another social more. You see, most New Yorkers of the day did not believe in the death penalty, especially in cases involving women, no matter what their reputation or social standing. And, in a strategic move right out of a rerun of *Law and Order*, Mary Alice wore mourning clothes every day in court. (Ah ha, the thick plottens!) Her wardrobe pandered to the Victorian sympathies of the judiciary. How could a bereaved woman—however wanton—be guilty of murdering her own mother?

To the shock of many, the sensational trial eventually resulted in an acquittal for Mary Alice (oh, those sly lawyers). She walked free, perhaps with just the hint of a Cheshire grin on her face.

Mary Alice continued to live in Manhattan, impoverished to the end, until her death (from natural causes, I might add) in 1948. Intentionally or otherwise, she may have ensured the reputation of clam chowder, if not as a preferred murder weapon, then, most certainly as an object (a recipe) of our attention and respect.

The recipes in this chapter are also to die for, although I shall assume, dear reader, that you will take that statement figuratively, not literally. Please!

Before We Begin—Some Clam Basics

Storing Fresh Clams

- Keep fresh clams cold in the refrigerator. Do not store in either freshwater or saltwater, as it will kill them.
- Store at a constant 41°F in the refrigerator in a container with the lid slightly open. Fresh clams are alive and need to breathe. Never put live clams in an airtight container. Drain any excess liquid daily.
- Clams will keep for at least four days at refrigerated temperatures.
- Live clams should close tightly when the shell is tapped. Discard any clams that do not close.

Preparing Fresh Clams

- Rinse the clams in cold running water.
- Before using, be sure the shells are closed. If some of the clams are slightly open, tap on them. If they slowly close, even partially, they are alive. If they don't close, they are no longer alive and should be discarded.
- Discard any clams with cracked or broken shells.

Cleaning Clams

- Scrub the clams with a stiff brush under cold running water.
- To remove the sand, put them in a sink or large pan of freshwater containing about half a cup of salt for each gallon of water. Leave hard-shell clams in this bath for about half an hour, then change the salted water and repeat the process twice. With soft-shell clams, make up the same water mixture and then sprinkle a little cornmeal on the water. This will help the clams purge any sand inside them.
- Refrigerate the clams in the brine mixture for at least four hours.

Clam Equivalencies

| 1 cup of clams | = one 6½-ounce can of clams |
| 2 cups of clams | = three 6½-ounce cans of clams |

1 quart of clams in the shell	=	2 pounds of clams in the shell
	=	1 cup of shucked clams plus 1 cup of liquid

Number of fresh clams in one quart:

Hard-Shell	Soft-Shell
18–24 Littleneck	18–24 Steamer
10–14 Cherrystone	10–18 Fry/Chowder
4–6 Quahog	

Canned Clams

Several of the recipes in this section use canned clams. I certainly realize that some individuals may wince at the mere suggestion of canned clams. (Trust me, I feel your pain!) However, there are several advantages of using canned clams in your recipes. They are available in almost every grocery store, large or small; I can always find them in my local semi-rural grocery store. For those readers who live in a land-locked state or far from any seashore, canned clams offer a most suitable alternative to fresh clams from seaside purveyors. Not only can canned clams be obtained throughout the year, they will also keep for long periods of time.

One of the most popular brand names for canned clams is "Snow's by Bumble Bee" (in their distinctive yellow-and-blue cans). You can find canned clams alongside canned tuna or in the soup or spaghetti sauce section of your grocery store. They are usually sold in six-and-a-half-ounce cans. There are three different sizes of clams: chopped (three-quarter-inch pieces), minced (half-inch pieces), and whole baby clams.

Recipes

"Easy Shmeazy" Clams in Their Shells

If you're just starting your clam culinary career, you can't go wrong with the following three "Easy Shmeazy" clam recipes. These uncomplicated meals involve a single ingredient—clams. Once you purchase your clams, bring them home, place them on the barbeque, in the oven, or in the microwave, and in no time at all you'll have a delicious meal for you and your family. These "quickies" are also great for those times when unannounced quests (like your really weird brother-in-law) drop by and you need something easy to prepare. Grab some clams, fire up the grill, oven, or microwave, and dinner will be ready before you know it. Of course, each of these recipes can be served as appetizers as well as on more formal occasions. Enjoy!

Barbeque:
Place scrubbed and clean clams on a hot barbeque about four inches from the flames. Grill for approximately ten minutes, or until the shells open.

Oven:
Place cleaned and scrubbed clams on a baking sheet on the middle rack of the oven. Roast at 350°F for ten minutes, or until the shells open.

Microwave:
Arrange cleaned clams around the outer circle of an oven-safe glass pie plate; cover with plastic wrap. Pierce the plastic wrap once or twice (for vents). Cook two to five minutes, or until the shells open. Rotate dish halfway through cooking time.

Serve the cooked clams in their shells along with individual cups of melted butter or cocktail sauce. Eating with your fingers is perfectly acceptable here. Watch out—the shells will be hot.

Classic Steamers

Some readers may recall the early years of television, long before satellite dishes and Netflix. For me, one of the programs I enjoyed watching every week was Dragnet. *Every episode would feature Sergeant Joe Friday and his sidekick out solving the various crimes that plagued Los Angeles in the 1950s. They spent much of their time interviewing possible witnesses or potential criminals. One of the classic lines from the show was, "Just the facts, ma'am."*

Well, you could say that this recipe is a "Just the clams, ma'am" method. It is simple, basic, and very easy to prepare, but it's one every cook should have in his or her repertoire. You'll be able to create dazzling appetizers or wonderful meals with just this recipe alone. Pair the clams with a good bottle of white wine and you could have a very romantic dinner.

1 tablespoon cornmeal
3 pounds steamer clams
1 stick butter, melted

1. Fill a large bowl or pot half-full with cold water. Sprinkle the cornmeal on the surface of the water. Scrub and clean the clams and place them in the cold water for at least two hours, changing the water (and cornmeal) two to three times. Drain the clams and rinse them well.

2. Put half an inch of water in the bottom of a large pot. Place the clams in the pot, cover, and turn the heat to high. When the water begins to boil, reduce the heat and uncover. Return the heat to high and cook uncovered until it boils again. Repeat alternating low and high heat about four more times, until the shells are completely open and the bellies look firm. This should take about ten minutes. Discard any clams that do not open.

3. Use a slotted spoon to transfer the clams to a large serving bowl. Pour the cooking liquid into several heat-resistant glasses and let the grit settle to the bottom. When the liquid has cooled sufficiently, carefully pour the broth into serving cups, leaving the grit behind.

4. While the clams are cooking, place the stick of butter in a small pan and warm over a low heat. Stir frequently until melted. Pour the melted butter into another set of serving cups.

5. Place a tall stack of napkins in the middle of the table and invite all your guests or family members to each grab a clam, remove the critter from its shell, dunk it in a serving cup of butter or clam juice, and enjoy. Forks (and good manners) are optional here.

Yield: 2 to 4 servings (as an appetizer)

Fried Clams

You could say that fried clams were nothing more than shelled bivalves cooked in very hot oil. But, that would be like saying that filet mignon is a piece of meat. Fried clams are to New England what barbeque is to Memphis, what pizza is to Chicago, what jambalaya is to Louisiana, and what Rocky Mountain oysters are to Colorado. Fried clams are inexorably tied to New England as much as any single food could be. As the legend goes, the first fried clams were created by the legendary "Cubby" Woodman on July 3, 1916, when he happened to put some breaded clams in one of the deep fryers used for cooking potato chips. The result was a revolution—one that has been going on for 100 years.

You can create that revolution right in your own kitchen with this all-purpose recipe. Prepare it as a summer treat for the whole family and they (just like tourists to New England) will keep coming back for more.

2½ pints shucked medium-sized, whole-belly soft-shell clams (canned clams are permissible)
3–4 inches of vegetable oil in a deep fryer
1½ cups evaporated milk
1½ cups yellow corn flour (see note below)
¾ cup all-purpose flour
Lemon wedges
Tartar sauce

1. Scrub and clean the clams. Open the shells and remove the raw clams, placing them into a bowl lined with paper towels. Or, if you are using canned clams, open the cans, drain the clams in a colander, and pat dry with paper towels.
2. Pour the vegetable oil into a deep fryer or heavy pot and heat over medium heat until it reaches a temperature of approximately 350°F.
3. Pour the evaporated milk into a medium bowl. In a large bowl, mix together the corn flour and all-purpose flour.
4. Using your hands, dip about one-third of the clams into the milk, letting the excess liquid drain off. Roll the raw clams in the flour mixture until each one is completely coated. Transfer the breaded clams to a colander. Shake the colander gently to remove any excess flour.
5. Using a slotted metal spoon, slide a spoonful of clams into the hot oil. Deep-fry the clams until they are a golden brown, usually two to four minutes, depending on the size of the clams.
6. After the clams are thoroughly cooked, place them in a Pyrex® dish in a warm oven (low heat) so they will stay hot while you prepare the remainder of the clams.
7. Serve with tartar sauce and lemon wedges.

Yield: 4 servings

NOTE: Make sure you use corn flour, not cornmeal. Corn flour is actually finely ground cornmeal. (Of course, you could always regrind your corn meal to get your own variety of corn flour.) Some "breading mixes" (depending on the brand) may be pure corn flour or corn flour plus salt (you want the unsalted corn flour). Read the labels carefully. If your local grocery store doesn't have corn flour in the baking section, check the Mexican foods section for *masa harina*, which is a good substitute. As a last resort, you can always order your corn flour online.

Roasted Clams

Summertime is the time for barbeques. Most people barbeque the standard items: hamburgers, hot dogs, chicken, fish, and steak. But if you're looking to spice up your next party or family gathering, you can't go wrong with roasted clams. They're easy to cook (even the family teenager can do it without excessive instructions), and they will quickly become the talk of your next barbeque. Serve them as an appetizer and you may never get around to the main course—they're that good!

Two dozen littleneck clams, scrubbed clean
Bottled cocktail sauce
Stick of butter
Lemon wedges

1. Fire up the barbeque (either a gas or charcoal grill). Wait until it gets to a moderately hot temperature.
2. Place each of the unopened clams on the grill; you may want to use kitchen tongs. Cook the clams for about five to ten minutes, or until most of the clams begin to open. Discard any unopened clams.
3. Use heavy gloves and pull off the top shell from each clam and discard. Dab about ½ teaspoon of cocktail sauce on top of each clam, then top off each clam with a thin slice of butter. (Note: You'll have to work fast here.)
4. Continuing to work quickly, return the clams to the grill (clam side up, shell side down) and continue to cook with the barbeque cover open until the clams are hot and the tops appear to be slightly glazed, about five minutes longer.
5. Remove the clams from the grill with the tongs and gloves, and place six of them on each of four small plates. Again, speed is of the essence here. Add a few lemon wedges to each plate and serve immediately. Put

a small bowl of extra cocktail sauce in the center of the table for dipping. Watch all the smiles.

Yield: 4 servings

Clam Fritters

Travel to any shore town on the East Coast during the summer months and you'll come across numerous vendors selling clam fritters. Some of those fritters are excellent, while others should probably be given to the local fishermen to use as sinkers. Suffice it to say, clam fritters are a staple of shore life—but they can also come in many different permutations and, certainly, a wide variety of tastes.

I've been searching for a yummy clam fritter recipe for many years and recently found this one. It's quick, easy, and will garner lots of rave reviews. All the ingredients can be found in every grocery store, and the preparation time is less than ten minutes. Enjoy!

2 (6½ oz.) cans minced clams
3 tablespoons (+ more for frying) vegetable oil
1 egg
¼ cup whole milk
1½ cups of all-purpose flour
2 teaspoons baking powder
¼ teaspoon salt
⅛ teaspoon pepper
Malt vinegar, lemon slices, Tabasco sauce

1. Pour the opened cans of minced clams into a colander resting over a bowl. Drain the clams and save the juice.
2. Pour about two inches of vegetable oil into a tall cup or medium bowl. Set aside.
3. In another medium bowl, whisk together the three tablespoons of vegetable oil and the egg. When completely combined, whisk in the milk and approximately ¾ cup of the reserved clam juice. Set this mixture aside.
4. In a large mixing bowl stir together the flour, baking powder, salt, and pepper. Add the egg mixture and then stir in the clams. (Note: Plan to adjust the batter so that it is the consistency of thick cake batter. You

can do this by adding a little more clam liquid or a little more flour as necessary.)

5. Pour about two inches of vegetable oil in a deep skillet. Heat to approximately 370°F. Quickly dip a long-handled spoon (an iced tea spoon works well) into the tall cup or medium bowl of vegetable oil to coat the spoon and prevent sticking. Dip out a spoonful of batter, keeping in mind that the size of the spoon will determine the size of your fritters. Place the batter into the hot oil and cook for about two to three minutes, turning once with the spoon. The cooked fritter should be golden and puffed on all sides. Remove and drain the fritters on paper towels.

6. It is strongly suggested that you fry the fritters in small batches (two to three at a time) until all the batter has been used. Cooked fritters can be kept in a Pyrex dish in a warm oven (about 200°F). If you are serving a large crowd, multiply the recipe amounts above by two or three.

7. Serve the fritters on individual plates and provide malt vinegar, lemon wedges, or Tabasco sauce for additional seasoning or dipping.

Yield: 6 servings of about 6 fritters each

Baked Stuffed Clams

I prefer recipes that don't require advanced degrees in chemical engineering or quantum mechanics. Recipes with scores of ingredients and dozens of directions immediately turn me off, most likely because I don't have the time to decipher what the writer is trying to say, or simply because I don't have six or seven hours to spend in the kitchen fixing the darn thing. This recipe, on the other hand, breathes simplicity. It can be on the table in less than half an hour, and it will generate compliments and kudos from young and old alike. It's also a recipe that can be modified and adapted depending on your culinary preferences; substitute new ingredients, add some unusual spices, or subtract an item just to see how it affects the taste. You'll discover all sorts of gastronomic possibilities.

18 clams (or 1–2 cans of minced clams)
1 tablespoon olive oil
2 tablespoons minced onion
1 clove garlic, minced
6 tablespoons breadcrumbs
2 tablespoons minced cured ham
¼ teaspoon lemon juice
Salt and pepper

¼ teaspoon paprika
1 tablespoon minced parsley
Butter

1. Preheat the oven to 350 ºF. Wash and scrub all the clams. Open each one carefully with a knife and remove the meat. Put all the meat on a cutting board and chop it up. (You may wish to consider using canned minced clams along with leftover clamshells from a previous recipe.) Reserve half the shells.
2. In a small frying pan, heat the olive oil. Sauté the onion and garlic until the onion is slightly wilted, about two minutes.
3. Stir in the breadcrumbs, ham, sherry, lemon juice, a little salt and pepper, paprika, and parsley. Combine thoroughly.
4. Add the clam meat to the mixture in the pan; adjust the amount according to taste. Fill each of the reserved shells with approximately one teaspoon of the filling.
5. Dot the filling in each clamshell with butter. Place the stuffed clamshells (stuffed side up) on a cookie sheet and bake in the oven for about ten minutes, or until slightly browned.
6. Remove the baked clams from the oven and serve on individual plates. Note that this recipe can be used as either an appetizer or as a main course (adjust the measurements above accordingly).

Yield: 4 to 6 servings

Down East Clam Pie

Typically associated with the folks who live on Long Island, clam pie is generally made of chopped or ground clams, onions, potatoes, spices, and cream or milk. To many people, clam pie is just like clam chowder with a piecrust on top. According to legend, this dish most likely came about as a simple, filling way of using up easy-to-harvest local clams and feeding the family for a day or two. As a result, it is a dish of frugality! You won't find this delight in many restaurants or in your local supermarket, but it's one that truly takes advantage of an excess of clams as well as lots of other items often found in the back reaches of the vegetable crisper.

¼ cup minced yellow onion
¼ cup minced red bell pepper
2 tablespoons minced parsley

6 tablespoons unsalted butter, separated
2–3 6½ oz. cans of chopped clams
24 saltine crackers
¼ cup clam juice
¾ cup milk or half-and-half
2 large eggs, lightly beaten
Salt and freshly ground black pepper, to taste
¼ to ½ teaspoon ground red chili pepper, optional

1. Preheat the oven to 325°F.
2. In a large sauté pan, combine the onion, red pepper, and parsley with two tablespoons of butter. Cook gently over medium-low heat until the vegetables are very soft, but not brown.
3. Pour the mixture into a large mixing bowl. Add the chopped clams.
4. Crumble half of the saltines between your hands and add them to the bowl with the clam mixture. Stir thoroughly.
5. Measure ¼ cup of clam juice and add it to the milk or half-and-half. You should now have one cup of liquid.
6. Pour the liquid mixture into the clam mixture. Add the eggs and stir thoroughly to mix all ingredients together. Add salt, black pepper, and chili pepper to taste.
7. Butter the bottom and sides of a 1-quart baking dish with two to three tablespoons of butter. Spoon the clam mixture into the dish.
8. Crumble the remaining saltines between two sheets of wax paper and use a rolling pin to crush them thoroughly. In a sauté pan, combine the finely crushed cracker crumbs with the remaining butter. Cook this mixture over medium heat until the crumbs just begin to brown.
9. Spoon the butter and crumb mixture evenly over the clam mixture. Place the dish into the oven and cook for one hour, checking after about 45 minutes. If the top of the casserole is still pale, turn the oven up to 425°F for the last fifteen minutes. If the recipe is starting to get too brown, place a sheet of aluminum foil loosely over top while the casserole finishes cooking. Remove the dish from the oven and serve immediately.

Yield: 4 to 6 servings.

New England Clam Chowder

The history of clam chowder goes back to the sixteenth and seventeenth centuries. The word chowder has its roots in the Latin word calderia, *which once meant "a place for warming things." Later, that same word was used to describe the implement that held those things—a cooking pot.*

For many people hundreds of years ago, a chowder was simply a poor man's food, a dish that brought together some combination of vegetables, seafood, venison, beef, pork, or anything else that wasn't nailed down into a stewy mixture that never had the same taste twice. Clams were a popular ingredient simply because they were so easy to obtain, particularly in New England. Although there are about a thousand different permutations of the basic recipe, this one stands as an excellent example of what constitutes a basic New England clam chowder, one with a milky consistency and the two essential ingredients—bacon and clams.

3 8-oz. bottles of clam juice
1 pound russet potatoes, peeled and cut into ½-inch pieces
2 tablespoons butter
3 slices bacon, finely chopped
2 cups chopped onions
1¼ cups chopped celery with leaves (about 2 large stalks)
2 garlic cloves, chopped
1 bay leaf
¼ cup all-purpose flour
6 6½-oz. cans chopped clams, drained, juices reserved
1¼ cups of half and half
1 teaspoon Tabasco sauce
Salt and pepper

1. Pour the clam juice into a large, heavy saucepan. Add the potatoes and bring to a boil over high heat. Immediately reduce the heat to medium-low, cover, and simmer until the potatoes are tender, about ten minutes. Remove the saucepan from the heat.
2. Melt the butter in a separate large pot over medium heat. Add the bacon and cook until it begins to turn brown, about eight minutes.
3. Add the onions, celery, garlic, and bay leaf to the bacon and sauté until all the vegetables soften, about six minutes.
4. Stir in the flour and cook an additional two minutes, stirring continuously so that the flour does not brown.

5. Add the potato mixture, the clams, the half-and-half, and the Tabasco sauce to the mixture.
6. Simmer the chowder for five minutes to blend all the flavors together. Stir frequently. Season to taste with salt and pepper.

NOTE: You can improve the taste of the chowder by preparing the recipe one day in advance. Refrigerate uncovered until it gets cold. Then, cover and keep refrigerated overnight. When ready to serve, put the pot on the stove and bring to a simmer. You'll note that this extra "seasoning time" enables all the flavors to meld together into a beautiful and tasty medley.

Yield: 4 to 6 servings

Manhattan Clam Chowder

Leave it to New Yorkers to do something different. Never satisfied with the status quo, the folks in the Big Apple have created their own distinctive version of clam chowder, one that has achieved its own popularity throughout the country.

This particular style of clam chowder has added more fuel to the "chowder wars"—that is, which style of chowder is the best? Walk into any restaurant north of Massachusetts and you could literally incite a riot by stating that the chowder from Manhattan (a place at least 300 miles away) is the best. Not only will that chowder have different ingredients, it will also have diehard supporters. A word to the wise: don't mention Manhattan clam chowder anywhere in Maine. You'll be lucky to escape with your life!

Nevertheless, New Yorkers persist in their diehard devotion to their clam chowder. The one ingredient that distinguishes this chowder from other varieties is the addition of tomatoes. The following recipe features a rich tomato taste. Do as I do, and prepare the chowder the day before its intended use. It truly improves with age.

1 tablespoon butter
2 slices bacon, diced
⅔ cup diced onion
½ cup diced celery
2 tablespoons minced green bell pepper
1 clove garlic, finely minced
1 cup peeled and cubed potatoes
1 teaspoon salt
¼ teaspoon black pepper, freshly ground

1 bay leaf
2 cups boiling water
2 cups stewed or canned tomatoes
1 tablespoon ketchup
2 cups clams, chopped fine, with liquid
¼ teaspoon dried thyme
⅛ teaspoon cayenne pepper
1 teaspoon minced parsley

1. Melt the butter in a large Dutch oven. Add the diced bacon and fry until the fat melts and the bacon is crisp and golden. Use a slotted spoon to remove the bacon and place it on several paper towels.
2. Put the onion, celery, bell pepper, and garlic into the Dutch oven. Sauté until onion is transparent, about two to three minutes.
3. Add the potatoes, salt, pepper, bay leaf, and boiling water. Bring to a boil, then simmer for about ten minutes.
4. Add tomatoes and ketchup to the mixture. Continue to simmer until all the potatoes are soft.
5. Add the clams, thyme, cayenne pepper, parsley, and saved bacon bits.
6. Simmer for an additional three minutes. Remove the bay leaf and discard. Serve in large bowls.

NOTE: You can adjust the measurements above. This recipe is sure to generate lots of compliments, so you might want to "double up" or even "triple up" the measurements.

Yield: Approximately 4 servings

Linguine with Clams

Ah, linguine! According to culinary historians, this classic Italian pasta got its start in Genoa and the Liguria region of Italy. It is what could be termed a "mid-sized" pasta—slightly broader than spaghetti, but not quite as wide as fettuccine. (By definition, linguine *means "little tongues" in Italian.) But, ah, what flavor! It is most often paired with seafood dishes, and it has become a common compliment to clam recipes. The recipe that follows is one of my all-time favorites, not only because it is quite simple to make, but also because it will generate lots of compliments. Be prepared to offer seconds and thirds.*

12 ounces of linguine pasta
2 tablespoons extra-virgin olive oil
2 tablespoons unsalted butter
3 cloves of garlic, minced
2 6½-oz. cans minced clams, drained and juices reserved
½ cup of heavy cream
Salt and pepper to taste
Grated Parmesan cheese
Parsley flakes

1. Bring a large pot of water to boil. You may wish to add a pinch of salt to the water. Cook the linguine according to the package directions until it is *al dente*. Pour the cooked linguine into a colander and allow to drain thoroughly.
2. Meanwhile, in a saucepan or large skillet, heat the olive oil and butter over medium-high heat until the butter is completely melted. Add the minced garlic to the pan and sauté until golden and fragrant, about one minute.
3. Add the reserved clam juice to the pan. Bring the mixture to a simmer and reduce by about half. With the heat on medium-low, stir in the clams and the heavy cream. Add salt and pepper to taste.
4. Ladle the linguine into individual serving bowls. Spoon some of the sauce over the pasta in each bowl. Top each bowl with freshly grated Parmesan cheese and parsley flakes. Serve immediately.

Yield: 4 servings

Clams and Chorizo

Not too long ago my wife and I went with some friends to one of our favorite restaurants in town. For my appetizer I ordered "Clams and Chorizo" simply because I'm always looking for innovative clam recipes and had never tasted this unique combination before. Well, let me tell you, my taste buds were doing a tap dance inside my mouth. The combination of clam and chorizo was a match unlike any other. Simply put, this was so good that I wanted to cancel my main course and place five more orders for this appetizer.

But, you don't have to use it as an appetizer. It works equally well as a main course with sourdough bread and plates of olive oil for dipping. One taste and your taste buds will be doing little dances inside your mouth, too.

2 tablespoons olive oil
½ pound Spanish chorizo sausage, diced
1 large fennel bulb, diced
1 large Spanish onion, diced
4 cloves garlic, grated or minced
1 teaspoon pimentón (paprika may be substituted)
Pinch of crushed red pepper
2 cups clam juice
1 cup dry white wine
3 dozen littleneck clams, scrubbed clean
4 tablespoons roughly chopped parsley
2 tablespoons unsalted butter
1 lemon, zested and juiced
Loaf of sourdough bread

1. In a large sauté pan (with lid) heat the olive oil over medium-high heat. Add the chorizo, cooking until it has let off some of its fat and the oil is colored by its spices, about three to five minutes.
2. Add the fennel and onion and cook until softened and fragrant, about four minutes. Next, add the garlic, pimentón, and crushed red pepper and cook for one additional minute.
3. Add the clam juice and white wine and bring the mixture to a boil. Quickly add the clams into the broth. Cover and cook until the clams open up, about six to ten minutes. Discard any clams that don't open.
4. After the clams have opened, add the parsley, butter, lemon zest, and juice. Ladle the clams, along with plenty of the liquid, into large bowls. Provide a loaf (or two) of sourdough bread, cut into fat slices, to sop up the delicious broth.

NOTE: Pimentón is a crucial ingredient in lots of Spanish dishes. It is often used to add color and flavor to many dishes from the Mediterranean region. It may be sweet (most popular), spicy, or smoked. You can usually find it in the international section of your local grocery store. If not, it can be easily obtained online.

Yield: 4 to 6 servings

Clam Scampi with Fettuccine

For culinary purists, the word "scampi" refers to either lobster or shrimp. In fact, if I had the space, I would share with you my all-time favorite recipe for shrimp scampi. But, since this is a book about clams, we're going to use the alternate definition for scampi—a style of preparation, rather than simply a primary ingredient. As a result, it's possible to discover dishes like chicken scampi (and others) on restaurant menus throughout the country. As you might suspect, scampi has both geographical and regional interpretations. Nevertheless, the one ingredient essential to all scampis is garlic. Without garlic there is no scampi, no matter what the main ingredient may be.

4 tablespoons butter
2 tablespoons olive oil
4 garlic cloves, minced
2 pounds hard clams, rinsed well
½ cup chopped fresh parsley
2 teaspoons lemon juice
Salt and pepper to taste
½ cup dry white wine
¼ cup sun-dried tomatoes
1 large tomato, diced
8 ounces spinach fettuccine, cooked

1. Put the butter and olive oil in a large skillet and heat over medium heat.
2. Add the garlic and continue to cook and stir for an additional minute.
3. Add the clams, stirring and cooking for five more minutes.
4. Add the parsley, lemon juice, salt, pepper, and wine. Cook for two more minutes, stirring occasionally, until the clams open. Discard any unopened clams.
5. Stir the tomatoes into the mixture and serve over bowls of fettuccine.

Yield: 4 servings

Clams Casino

Here's another "must-have" clam recipe that is rooted in both legend and innovation. According to the story, the initial recipe for Clams Casino was developed in 1917 in the Little Casino in Narragansett, Rhode Island. A certain "woman of means" wanted to impress her guests, so she asked the chef to create something original. The new recipe was an overwhelming success and was eventually named for the hotel. Word of its success soon spread across the country where the basic recipe has been modified and altered according to local customs or inclinations. In New Orleans, for example, the clams have been replaced by oysters (certainly a culinary faux pas in my book). A staple of most Italian restaurants, the one constant in every recipe is the bacon.

8 slices lean bacon, chopped fine
1 cup chopped onion
2 large garlic cloves, minced
1 cup finely diced red bell pepper
1 cup finely diced green bell pepper
½ teaspoon dried oregano, crumbled
2 tablespoons olive oil
2 teaspoons wine vinegar
2 tablespoons freshly grated Parmesan cheese
Salt and pepper to taste
2 dozen middleneck clams, shucked, bottom shells reserved
Rock salt for lining the pan and platter

1. Preheat the oven to 400°F.
2. In a heavy skillet, cook the chopped bacon over medium heat until browned. Remove the bacon to some paper towels to drain. Wipe the skillet clean.
3. Put the onion, garlic, bell peppers, and oregano in the skillet with the olive oil. Cook over a low heat until peppers are crisp, yet tender. Now, transfer the mixture to a small bowl.
4. Stir in the chopped bacon, vinegar, and Parmesan cheese. Add salt and black pepper to taste.
5. Fill a large flat pan with a layer of rock salt (to balance the shells). Place individual clams shells face up in the pan. Divide the clam meat equally among all the shells. Top each clamshell with the mixture of onion, bacon, and pepper.

6. Bake in the oven for twelve to fifteen minutes, until cooked through. Serve clams on a fancy plate lined with a layer of rock salt (again for balance).

Yield: 4 servings (as an appetizer)

Clams Provençal

You know this is a classic French recipe when you see the ratio of wine to all the other ingredients. But it's not the wine that is important; rather, it is the intricate blending of all the ingredients into a dish that may quickly silence an entire dinner table . . . with the exception of all the o-h-h-h-hs and a-h-h-h-hs echoing around the dining room. This is a dish that will be remembered long after your guests leave your house and the last plate has been scrapped and the dishwasher turned on. This is a dish that will generate wide smiles . . . and lots of repeat "customers."

½ cup virgin olive oil
1 onion, finely chopped
1 red bell pepper, diced
4 vine-ripened tomatoes, diced
½ stalk celery, sliced
2 cloves garlic
2 pounds clams, cleaned
⅔ cup dry white wine
⅓ tablespoon chopped thyme
⅓ tablespoon chopped rosemary
⅓ tablespoon chopped marjoram
Salt and pepper to taste

1. Place the oil, onion, bell pepper, tomatoes, celery, and garlic in a large cooking pot. Cook over high heat for five minutes, stirring frequently to prevent sticking.
2. Add the clams, white wine, fresh herbs, and salt and pepper. Cook with the lid on until all the shells have opened, about eight to ten minutes. Stir frequently to ensure even cooking. Discard any unopened clams.
3. Ladle into large bowls. Serve with an oil and vinegar salad and a grilled sliced baguette. Don't forget a bottle of rosé or white wine.

Yield: 4 to 6 servings

Acknowledgments

SUPERMAN. BATMAN. WONDER WOMAN. SPIDERMAN. Superwoman. The Flash. Aquaman. Those were my heroes and heroines when I was growing up. I looked up to them for inspiration, guidance, and a moral code. If I could be like them, I thought to myself, then I would have a most wondrous (and adventurous) life. Not only would I be able "to leap tall buildings in a single bound," but I'd also be able to vanquish any and all foes, save the world from destruction by demonic tyrants, and preserve the American way of life forever. Wow, what a legacy!

As a writer I still have heroes—especially all those folks who unselfishly gave of their time and talent to help me on this incredible journey of discovery and adventure. These are the people who made this book much more than the work of the guy whose name appears on the cover; these are the individuals who shared their talents, their time, and their expertise to ensure that the story of clams is told accurately, thoroughly, and well. I am indebted to each and every one of them for holding my literary hand and guiding me down new paths and through unknown vistas to learn about one of the most amazing creatures on the planet.

To Cindy Bear and Jim Fors of the Randell Research Center in Pineland, Florida, who gave me both perspective and education on the role of shell middens as archeological timelines, I offer a hearty (and standing) round of applause.

I would like to extend a plethora of high fives and a chorus of cheers to Jon Gill and Shawn Stephenson of Southern Cross Sea Farms, who provided me with an up-close-and-personal introduction to the remarkable world of clam aquaculture. They are truly modern-day Aquamen!

In my travels I encountered many wonderful people, and Alice Phillips, proprietor of the lovely and charming Cedar Key Bed & Breakfast in Cedar Key, Florida, certainly stands at the top of that list. She manages one of the finest hostelries in the country—filled with warmth, good conversation, and the famous "bottomless" cookie jar. She is, indeed, a class act!

My sincerest thanks and appreciation to Tammy and Todd Smith, also of Cedar Key, who offered me delightful insights and personal stories about the life and times of professional clammers. I enjoyed their camaraderie as much as I did their constant and positive outlook on life.

I am incredibly indebted to Rick Bushnell of Re-Clam the Bay, who gave me a guided tour of environmental stewardship and the passionate volunteerism endemic to Barnegat Bay, New Jersey. He deserves every accolade possible for his interest and support of this book and for his role in preserving our favorite bivalves.

I am especially indebted to one of my colleagues in the Department of Education at York College of Pennsylvania, Irene Altland. Not only does she excel at deciphering my mumblings on ancient cassette tapes, so, too, does she produce meaningful and coherent transcripts of my travels and interviews. She gets lot of hugs and, of course, lots more work in the future.

To the Supermen of computer technicians, Joseph Kiel and Mark Bierman of Fast Teks Computer Services (www.fastteks.com/centralmasondixon), who saved my life (and significantly lowered my blood pressure) when my computer "crashed and burned" two months before this manuscript was due to my editor, I extend my undying and everlasting appreciation. Through technological magic that I will never understand, they were able to painstakingly retrieve all my notes, drafts, and final documents from the far reaches of cyberspace and helped make this book a (timely) reality. My neighbors are equally appreciative of their services, since I am no longer shrieking and screaming into the wee hours of the night.

To Marco Pinchot and Karen Underwood of Taylor Shellfish, who gave me tours of some incredibly magical places and delightful vistas, I am forever grateful. Their passion for clams and their dedication to the mission of a celebrated company are to be both admired and cheered. Clair and Brittany also deserve enormous thanks for their contributions and anecdotes.

No project of this magnitude happens without an excellent editor . . . and Lindsey Breuer was both guide and sage on this literary venture. She brought coherence to my musings, story to my digressions, and consistency to my journalistic wanderings. She was an absolute joy to work with and a professional of the highest caliber.

And, finally, to all the numerous unnamed people I met as I crisscrossed the country from New England to California and from Washington State to Florida—folks who shared an ancient story, taught me how to use a clam tube, proffered a family recipe, pointed out a historical document, instructed me on the finer points of clam physiology, told me about a life on the sea, educated me about clam sex (it's not pretty!), or simply served up some of the best fried clam strips in the universe—I thank you enormously. I never got all your names, but I did catch your spirit, your energy, and your passion for the most remarkable bivalve known to exist. Thank you one and all for your time, talents, and personal investments in this tale.

Clams That Fly

She ate so many clams that her stomach rose and fell with the tide.
—Louis Kronenberger

ET'S SAY YOU LIVE IN CARBONDALE, COLORADO, or Dover, Pennsylvania. You wake up one day with an overwhelming desire for clams, but you soon realize that you're not quite in the right place to gather clams in your backyard. (Actually, you are far *far* away from a convenient ocean.) What to do?

Fortunately, there are a number of businesses that exist near the appropriate bodies of water who will be more than happy to speed those tender little delicacies to you so they will be on your dinner table by the next day.

That's right, there are clams that fly!

The following purveyors and suppliers will gladly ship clams to you, no matter where you live. Most will harvest the clams the same day they are ordered so that you can be assured that when you receive them they will only be hours (not days) old. Please keep in mind, however, that, depending on the quantity you order, overnight shipping costs may exceed the actual costs of the clams themselves. But, hey, what's price when you're planning to prepare some of the best "Clam Scampi with Fettuccine" in the western hemisphere?

Take some time to browse the various websites listed below. You will quickly notice a wide variety of offerings and services. Since you are spending some serious dollars here, I would suggest you call two or three of the companies (closest to where you live) to inquire about the quality of their product(s) as well as their shipping procedures. Here are a few questions you may want to ask:

- How soon after I place my order will the product go out?
- What will be the exact cost of my order (product + shipping)?
- What carrier (FedEx, UPS, USPS, etc.) do you use to ship your products?
- Do you offer any guarantees along with your products?
- Under what conditions do you offer refunds?
- How, exactly, will my order be packed and wrapped (gel packs, Styrofoam, plastic wrap, ice bags, vacuum sealing)?
- Will someone need to be home to accept the package? Will someone need to sign for the package when it arrives?
- Do you provide any recipes or specific cooking instructions along with your products?
- Can you provide me with testimonials or endorsements of your products or services?

The following companies are listed in alphabetical order. Check out the websites of those within geographical proximity to where you live or with the specific products you desire.

Blue Crab Trading Company
Easton, MD
(800) 499-CRAB (2722)
www.bluecrabtrading.com
This vast mail order seafood company specializes in Virginia littleneck clams. If so inclined, you can also order alligator meat ("it tastes just like chicken") to compliment your clam dinner.

Browne Trading Company
Portland, ME
(800) 944-7848
www.brownetrading.com
This is one of the premier mail-order companies in New England. You can order littleneck clams harvested off Cape Cod in packages of fifty clams each.

Cape Porpoise Lobster Company

Cape Porpoise, ME

(800) 967-4268

www.capeporpoiselobster.com

Here is where you can get live Maine softshell steamer clams. Their prices also include overnight shipping.

Charleston Seafood

Charleston, SC

(800) 609-FISH (3474)

www.charlestonseafood.com

Charleston Seafood delivers fresh fish and seafood to all fifty states. In addition, they offer weekly specials, fish and shellfish recipes, seafood preparation and cooking tips, and gourmet gift baskets. Their littleneck clams are sold in bags of one hundred.

Copps Island Oysters

Norwalk, CT

(203) 866-7546

coppsislandoysters.com

The folks here will ship directly to your door within two hours of your order; place your order before 11:00 a.m. and it goes out the same day. They specialize in littlenecks, topnecks, and cherrystone clams.

Ed's Kasilof Seafoods

Kasilof, AK

(800) 982-2377

www.kasilofseafoods.com

Ed's Kasilof Seafoods is a family-owned and -operated gourmet seafood company. They have been selling wild Alaskan salmon and processing gourmet seafood since 1970. Check out their Alaska wild razor clams directly from the waters of Cook Inlet.

THE SECRET LIFE OF CLAMS

Fisherman's Fleet

Jonestown, ME

(888) 732-3663

fishermansfleet.com

This company offers a wide variety of seafood available for shipping all over the country. Choose from the following: cherrystones, chopped clams, cultured count neck clams, and fryer clams. The whole clams are priced by the clam, so you can order exactly the number you need.

The Fresh Lobster Company

Gloucester, MA

(508) 451-2467

www.thefreshlobstercompany.com

Headquartered in historic Gloucester, Massachusetts, this company can provide you with live Ipswich softshell clams, fresh littleneck clams, raw frying clams, raw chopped clams, large littleneck clams, and fresh clam juice.

Great Alaska Seafood

Soldotna, AK

(866) 262-8846

www.great-alaska-seafood.com

If you're in the market for fat, sweet and succulent razor clam steaks direct from Alaska, then this is your source.

Ipswich Fish Market

Ipswich, MA

(888) 711-3060

www.ipswichfishmarket.com

You'll love the offerings from this well-known seafood supplier. You can order the famous Ipswich clams (dug by hand), in addition to clam chowder, a ready-to-go New England clambake, a clambake express, a Clam Frying Kit with "The Clam Box" secret recipe, and shucked clams.

Legal Sea Foods

Cambridge, MA

(800) 343-5804

http://shop.legalseafoods.com

In business for more than six decades, Legal Sea Foods has an extensive offering of seafood by mail. You'll want to check out all the various types of clam chowder you can have delivered right to your front door.

Linton's Seafood

Crisfield, MD

(877) 546-8667

lintonsseafood.com/clams

Located on the southern tip of Chesapeake Bay, Linton's has been shipping fresh clams to customers for more than thirty-five years. They specialize in littlenecks.

The Lobster Guy

Point Judith, RI

(866) 788-0004

www.thelobsterguy.com

This company offers a wide array of seafood that can be shipped overnight to any continental US destination. Check out their fresh steamer clams and fresh littleneck clams, all of which are dug daily by local clam diggers. You can also order a complete clambake for two to six people.

Maine Lobster Direct

Portland, ME

(800) 556-2783

www.mainelobsterdirect.com

This enterprise specializes in Maine steamer clams as well as their own "damn good" clam chowder. Check out their stuffed clams, too.

MarxFoods

Seattle, Washington

(866) 588-6279

www.marxfoods.com

One of the largest clam distributors in the Northwest. You can obtain a wide variety of live clams, including geoducks, large quahogs, small quahogs, Manilas, steamers, and mahogany clams.

Patriot Lobster
Marblehead, MA
(866) 456-2782
www.patriotlobster.com
This company is a family-run operation that ships seafood products all over the world. Check them out for littleneck clams, cherrystone clams, quahog clams, and steamer clams. They ship overnight via FedEx.

Pike Place Fish Market
Seattle, WA
(800) 542-7732
www.pikeplacefish.com
This is perhaps the most famous fish market in the United States. (Yes, they actually throw fish around; I almost got sideswiped by a soaring salmon one day!) You can also use their mail-order services to get quality seafood delivered to your door; no, they won't throw it through your window. They offer live Manila clams as well as razor clam meat.

Pure Food Fish Market
Seattle, Washington
(800) 392-FISH (3474)
www.freshseafood.com
You can get Manila steamer clams and razor clams here. All orders are shipped overnight by FedEx.

Simply Lobsters.com
Lewiston, ME
(800) 796-3189
www.simplylobsters.com
This company sells Maine steamer clams by the pound. You can also place large orders (up to seventy pounds) for family reunions, tailgating parties, or other large functions.

South Beach Fish Market
South Beach, OR
(866) 816-7716
www.southbeachfishmarket.com
This Oregon-based operation ships live gourmet littleneck clams throughout the Pacific Northwest.

Taylor Shellfish Farms

Shelton, WA

(360) 432-3300

www.taylorshellfishstore.com

This company has been a family operation for more than one hundred years (see chapter 9). Here you can order manila clams as well as live geoducks (one of their bestsellers).

Trenton Bridge Lobster Pound

Trenton, ME

(207) 667-2977

trentonbridgelobster.com

This company specializes in steamer clams. The minimum order is five pounds.

Reading Group Guide

"Never hold discussions with the monkey
when the organ grinder is in the room."

—Winston Churchill

Gather a large group of friends together and declare, "Hey, guys, let's talk about clams" and, most likely, several folks will find an excuse to refill their cocktail glasses ("Make that a double!"), have an overwhelming compulsion to rotate the tires on their cars (like right now), or have a sudden urge to use the bathroom facilities for long periods of time. Quite obviously, the mere thought of an extended discussion on marine invertebrates would not have the cachet of, say, a deep and abiding conversation about the increasing stupidity of politicians, the latest antics of some Hollywood starlet (with a penchant for "wardrobe malfunctions"), or the current rage in diets, electronic doodads, or vacation escapes.

I'm well aware of the fact that clams, as a conversational topic, are not particularly sexy.

But let's change that equation for a moment. Let's assume you've read this book and your mind is still spinning from all the incredible information, delightful insights, and wonderful stories you devoured (intellectually speaking). You look, once again, at the cover of the book and say to yourself, "You know, this would make for a really cool discussion the next time we get the 'gang' together!" (Trust me, I won't hold it against you if you talk to yourself a lot . . . so do a lot of authors.) Then, you think, "You know, I could get everyone together and we could all cook some of those really neat clam recipes, eat ourselves silly (drinking yourself silly is optional), and then have a really intelligent discussion about clams for the rest of the night. How cool would that be?"

Actually, cooler than you could ever imagine!

If you really want to get the group started on a fascinating cognitive journey (archaeologically speaking), you might want to bring out your kitchen trashcan and empty it on the floor (place some old newspapers down first). Invite guests, in teams, to paw through your garbage (perhaps some latex gloves should be passed around) to make some guesses or assumptions about how you and your family live your lives. What can your guests deduce about your lifestyle? What are some notable habits? What type of diet do you and your family follow? What differences are there between your life and the lives of your guests? Encourage groups to share their insights and suppositions based entirely on the contents of your trashcan.*

Invite members of the group to each share the most interesting, fascinating, or surprising fact about clams. What was the most amazing piece of information discovered in the book? What was absolutely stunning about clams? What was something you learned that you found totally mind-blowing? Ask group members to defend their choice of "Fast Fact" and why they believe their choice is the most incredible. Record these on sheets of paper and then rank order them from "Most Amazing" to "Least Amazing." Post the results of your informal survey on Facebook or Pinterest. See if others agree with your listings.

If your guests are accomplished cooks (or just think they're accomplished cooks), invite each one to bring their favorite clam recipe. (Pilfering the recipes from this book is perfectly acceptable.) Make room in the kitchen for various individuals or cooks to prepare their recipes. If need be, create a cooking rotation—staggering the various dishes and chefs. Serve each dish to all the guests and invite them to comment on each one or to rank them from 5-star to 1-star (hopefully none of the latter). Perhaps some

---

* If you want to be really sneaky, this might be a good opportunity to place some select items in your trash before your guests arrive. Might I suggest some receipts from the fanciest restaurant in town, a fictitious letter from the president lauding your community service, or a discarded email from a certain magazine inviting you to pose for its annual swimsuit issue!

intrepid soul would enjoy assembling all those recipes together into a mini-cookbook to share with group members as a remembrance of the wild and wacky time they had while in your kitchen.

If you have several creative types in your group, provide various teams with sheets of paper and some writing implements. Have each group list the letters in "CLAM" vertically down the left side of a piece of paper. Give each group a time limit (five minutes, for example) and ask them to come up with a "clam adjective" or "clam phrase" for each of the letters. Each adjective must relate to information or data presented in the book. Here's mine:

C—*Cuddly? Not particularly.*
L—*Luscious with butter*
A—*Absolutely amazing, astonishing, and astounding animals*
M—*Millions of gametes; thousands of grandchildren*

After each team has designed its own acronym, invite the entire group to vote on the best. Which one was most descriptive? Most informative? Most humorous? Most obvious; least obvious? Post your acronyms online; solicit some reactions.

Okay, the intellectuals in the group are getting restless. They're starting to yawn, stretch, and mumble under their breath. They need something to keep their neurons firing, or they'll revolt en mass (revolting intellectuals—definitely not a pretty sight). Try this: Divide the group into several small teams and provide each team with two clamshells. Give the teams a set amount of time (and a refill on their drinks) and challenge them to come up with as many possible tools as they can think of that can be made from clamshells (similar to what ancient peoples had to do when faced with a plethora of excess shells left over from a family celebration or community festival). Here's a short list of possibilities I devised (you will quickly notice that I am definitely not a revolting intellectual . . . or even a regular intellectual, for that matter):

- Spoons (attach a wooden handle)
- Soup ladle (again, an extended handle is needed)
- Small digging tool (for small garden plants)
- Calcium earmuffs for ceramic figurines (I didn't say the tools had to be logical.)
- Nose ornament (a fashion statement by Lady Gaga)

- Stepping stones for squirrels (placed face down in a row in the garden)
- Tiny birdbaths for tiny birds (placed face up in the garden)
- A necklace (for use in tropical locations)
- Musical instrument (castanets, anyone?)
- Miniature hot tubs for a seaside doll house (Okay, it's time to retire.)

You might want to award prizes (a clamshell sculpture) for the team that comes up with the most uses for clamshells or the team that devises the most distinctive list of items. Either way, you're sure to generate lots of laughs, if not lots of pleasant memories with this activity.

In my "neck of the woods" it is not uncommon for groups of people to engage in what is affectionately known as "pub crawls"—dropping in on various taverns, pubs, and bars to sample the wares for a while and then move on to the next one on the list (not that any formal list is required for this activity). Particularly popular with college students around St. Patrick's Day (certainly not the students at my college!), these high-octane-fueled journeys make for many unforgettable memories (as well as many unforgettable hangovers the next morning). Depending on where you live, or where you are vacationing, you might want to assemble a group and do a "clam shack crawl"—a journey to as many clams shacks as possible (in a seventy-two-hour period, for example) to sample their fried clam strips, clam chowder, or steamed clams. Rate them, rank them, evaluate them. Which is the best overall? Which one has the best chowder? Which one has the best steamed clams?

After all is said and done, you'll probably discover that the seemingly sedate creatures known as clams are quite exciting after all.

Discussion Questions

Here are some questions to consider as you and your friends review the information in this book. Serve these up with your favorite clam recipe (and an equally favorite bottle of wine), and the conversation can go on all night.

1. What do you see as the biggest difference between people who love oysters and people who love clams? Is it possible to have "a happy medium?" How do you think most of the country stands on this "issue?" What would you like to say to "the other side?"

2. What would you consider to be the best clam recipe ever created? What ingredients were in that recipe that made it so good? What do you think is the best restaurant that serves clams? What distinguishes that place from all the other restaurants you've been to?

3. The author provided several examples of clams in our culture. In what other ways have clams entered our culture? In what ways have clams become part of the American experience?

4. What was the most fascinating aspect or information about clams? What did you learn from reading the book that stands out as memorable for you? In what ways do you see clams differently as a result of reading this book?

5. Clams have a long history . . . a very long history (say, more than 500 million years long). Why do you think they have been so evolutionarily successful? Why have they endured when so many other animal species have died out?Is there something about their anatomy, physiology, or lifestyle that has contributed to their longevity?

6. The book shared information on why clams are "nutritional super-stars." Will you make clams a more regular feature in your diet? How will you "sell" clams to other members of the family?

7. Let's say you were asked to develop a thirty-second Super Bowl com-mercial about clams—that is, you have been asked to inform the buy-ing public about the importance of including more clams in their lives. What will you include in your commercial that will help you achieve that goal and make clams a celebrated food (just like Anheuser-Busch made beer a celebrated beverage during the Super Bowl)?

8. How would you describe clams using just six words? Can you design a six-word phrase or sentence that encapsulates the significance or life story of clams? Here are a few to get you started:
   • Sand critters lost beneath rolling tides.
   • Succulent denizens crowding vast chowder seas.
   • Roasted, fried; steamed and baked. Yum!
   • Reproductive frenzy—billions of potential offspring.

9. Should clam aquaculturists increase their production of commercial clams? What will more clams on the market mean for Americans? Will most Americans ever embrace clams in the same way they embrace T-bone steaks and fried chicken?

10. What is the perfect drink or perfect cocktail to serve with clams? Red wine is often served with meat and white wine with chicken. But what should we serve with clams? If it's white wine, what would be the per-fect white wine to serve with clams? A white Bordeaux? Sauvignon blanc? California chardonnay? How does the method of clam prepa-ration (steamed, cream sauce, tomato sauce, etc.) affect the choice of wine?

11. What mixed drink goes best with clams? How about a clam martini (a "clamtini")?

> 3 oz. vodka
> ½ oz. dry vermouth
> 1 teaspoon clam juice
> Combine ingredients and shake with cracked ice. Pour into a martini glass and garnish with an olive.

> Or how about a "Seaside Firestarter?"
> 1½ oz. tequila
> 6 oz. Clamato juice
> ¾ tablespoon horseradish
> Splash of lemon juice
> Worcestershire sauce (to taste)
> Tabasco sauce (to taste)
> Mix all ingredients in a highball glass. Fill it with ice and garnish with a slice of lemon.
> (Warning: After consumption - do not breathe on anyone!)

Bibliography

Adams, Mike. "Aphrodisiac Claim for Oysters Finally Backed by Research." *Natural News*. March 26, 2005. Accessed October 15, 2013. http://www.naturalnews.com/z005962.html.

Aten, Lawrence E. "SHELL MIDDENS." *Handbook of Texas Online*. Published by the Texas State Historical Association. Last modified June 15, 2010. Accessed October 1, 2013. http://www.tshaonline.org/handbook/online/articles/bcs03.

Badger, Curtis J. *Clams: How to Find, Catch and Cook Them*. Mechanicsburg, PA: Stackpole Books, 2002.

BBC News. "Ming the Clam Is Oldest Mollusc." Last modified October 28, 2007. Accessed February 4, 2012. http://news.bbc.co.uk/2/hi/science/nature/7066389.stm.

Bloom, Harold. *The Western Canon: The Books and School of the Ages*. New York: Harcourt Brace & Company, 1994.

Brosius, Liz. "Fossil Clams and Other Bivalves." *GeoKansas*. Published by Kansas Geological Survey. Last modified June 15, 2005. Accessed October 14, 2013. d

Buttner, Joseph K. Scott Weston, and Brian F. Beal. "Softshell Clam Culture: Hatchery Phase, Broodstock Care through Seed Production." Fact sheet, Northeastern Regional Aquatic Center, NRAC Publication No. 202-2010. College Park, MD: University of Maryland Aquaculture Center, n.d.

Buzan, Tony, and Raymond Keene. *Buzan's Book of Genius (And How You Can Become One)*. London, England: Stanley Paul, 1994.

Caduto, Michael J., and Joseph Bruchac. *Keepers of the Earth: Native American Stories and Environmental Activities for Children*. Golden, CO: Fulcrum, 1997.

Carrol, Richard. *The Great Mussel and Clam Cookbook*. North Vancouver, British Columbia, Canada: Whitecap Books, 2004.

Cedar Key Aquaculture Association. *Cedar Key Everlasting*. Cedar Key, FL: Cedar Key Aquaculture Association, 2012.

Chesapeake Bay Program. "Soft Shell Clam." Accessed September 30, 2013. http://www.chesapeakebay.net/fieldguide/critter/soft_shell_clam.

David, J. "History of Clams." Accessed 1/25/2012. http://www.bukisa.com/articles/23549_history-of-clams.

DePratter, Chester B. "Georgetown County Marsh Middens and Clam Shell Analyses." *Legacy* 9, no. 3 (2005): 12-13.

"Difference Between Oysters and Clams." *DifferenceBetween.net*. Accessed October 12, 2013. http://www.differencebetween.net/science/nature/difference-between-oysters-and-clams.

Dojny, Brooke. *The New England Clam Shack Cookbook*. North Adams, MA: Storey Publishing, 2008.

Dunham, Will. "For Early Humans, A Beach Party and Clam Bake." *Reuters.com*. Accessed October 14, 2013. http://www.reuters.com/assets/print?aid=USN1732478620071017.

Ely, Eleanor. "Quahog." Rhode Island Sea Grant. Accessed September 30, 2013. http://www.seagrant.gso.uri.edu/factsheets/fsquahog.html (site discontinued).

Field Museum of Natural History. "Butter Clam." Accessed October 12, 2013. http://eol.org/pages/491722/details.

Fish and Wildlife Research Institute. "Clams: Florida's Buried Treasure." St. Petersburg, FL: Florida Fish and Wildlife Conservation Commission (June 2005).

Flagg, William. *The Clam Lover's Cookbook*. Gloucester, MA: North River Press, 1983.

Fleeson, Lucinda. "Mound of Evidence: A Heap of Oyster Shells May Provide Archaeologists with a Valuable Record Of How Man Lived In New Jersey In Prehistoric Times." *Philadelphia Inquirer*. April 17, 1995. Accessed October 1, 2013. http://articles.philly.com/1995-04-17/living/25688576_1_shells-stones-tool.

Flimlin, Gef. "Hard Clam Aquaculture in New Jersey." New Brunswick, NJ: Rutgers Cooperative Research and Extension (March 2003).

Fredericks, Anthony D. *Horseshoe Crab: Biography of a Survivor.* Washington, DC: Ruka Press, 2012.

———. *How Long Things Live: And How They Live as Long as They Do.* Mechanicsburg, PA: Stackpole Books, 2010.

———. *Walking with Dinosaurs: Rediscovering Colorado's Prehistoric Beasts.* Boulder, CO: Johnson Books, 2012.

Friedlander, Blaine. "Clam Fossils Divulge Secrets of Ecologic Stability." *Cornell Chronicle.* May 15, 2013. Accessed September 28, 2013. http://www.news. cornell.edu/stories/2013/05/clam-fossils-divulge-secrets-ecologic-stability.

Gelb, Michael J. *How to Think Like Leonardo da Vinci: Seven Steps to Genius Every Day.* New York: Bantam Dell, 1998.

"Geologist Researches Climate-Induced Downfall of Advanced Civilization." *Desktop News.* Last modified April 16, 2013. Accessed September 2, 2013. http://www.as.ua.edu/home/college-geologist-researches-evidence-of-ancient-climate-induced-downfall-in-clam-shells.

Go Ask Alice. "Aphrodisiacs – Do They Really Make You Randy?" Accessed October 18, 2013. http://goaskalice.columbia.edu/ aphrodisiacs-do-they-really-make-you-randy.

Goldman, Erica. "Clear Water through Clam Culture?" *Chesapeake Quarterly Online.* Accessed 10/12/2013. http://ww2.mdsg.umd.edu/ CQ/v06n2/side4/index.html.

Gould, Stephen Jay. *Leonardo's Mountain of Clams and the Diet of Worms: Essays on Natural History.* New York: Harmony, 1998.

Greenberg, Paul. *Four Fish: The Future of the Last Wild Food.* New York: Penguin Books, 2010.

Gugino, Sam. "Clams: A Shell Game Worth Playing." SamCooks.com. Accessed September 30, 2013. http://www.samcooks.com/food/sea-food/clams.htm.

Heilbrunn Timeline of Art History. "Winslow Homer: A Basket of Clams." The Metropolitan Museum of Art. Accessed October 4, 2013. http:// www.metmuseum.org/toah/works-of-art/1995.378.

Hellenica World. "Botticelli: The Birth of Venus." Accessed March 17, 2012. http://www.mlahanas.de/Greeks/Mythology/Master4.html.

Hirst, Kris. "The Archaeological Study of Shell Middens." About. com. Accessed October 1, 2013. http://archaeology.about.com/od/ boneandivory/a/shellmidden.htm.

Holmes, Judy. "Ancient Clams Yield New Information about Greenhouse Effect on Climate." *Inside SU.* Last modified August 18, 2011. Accessed March 18, 2012. http://insidesu.syr.edu/2011/08/18/ ancient-clams-yield-new-information.

Isabella, Jude. "The Edible Seascape." *Archaeology* 64, no. 5 (September/October 2011). Accessed March 19, 2012. http://archive.archaeology.org/1109/features/coast_salish_clam_gardens_salmon.html

j3nn, April 14, 2013 (07:51 a.m.), comment on "Clams vs. Oysters." Mark's Daily Apple. Accessed October 12, 2013. http://www.marksdailyapple.com/forum/thread83220.html.

Jacobsen, Rowan. *A Geography of Oysters: The Connoisseur's Guide to Oyster Eating in North America.* New York: Bloomsbury, 2007.

Jenkins, Nancy Harmon. "The Deep-Fried Truth About Ipswich Clams: No matter the source of the harvest, the secret to a classic seaside meal may be the mud." *New York Times*, August 21, 2002. Accessed October 11, 2013. http://www.nytimes.com/2002/08/21/dining/deep-fried-truth-about-ipswich-clams-no-matter-source-harvest-secret-classic.html.

Johnson, Kimberly. "New Giant Clam Found; May Have Fed Early Humans." *National Geographic News*, August 29, 2008. Accessed March 19, 2012. http://news.nationalgeographic.com/news/pf/9216133.html.

Jones, G. Stephen. "Clams." *The Reluctant* Gourmet (blog). Last modified August 9, 2012. Accessed October 12, 2013. http://www.reluctantgourmet.com/clams.

Julian, Sheryl. *The New Boston Globe Cookbook.* Guilford, CT: Globe Pequot Press, 2009.

Kipfer, Barbara Ann. "Results for Midden." Archaeology Wordsmith. Accessed November 22, 2013. http://archaeologywordsmith.com/lookup.php?terms=midden.

Kohls, L. Robert. *Survival Kit for Overseas Living.* Yarmouth, ME: Intercultural Press, Inc., 1996.

Kurlansky, Mark. *The Big Oyster: History on the Half Shell.* New York: Random House, 2006.

Living Australia, The. *Dangerous Australians: The Complete Guide to Australia's Most Deadly Creatures.* Sydney, Australia: Murdoch Books, 1986.

Mahr, Krista. "Top 10 Scientific Discoveries." *TIME*, December 9, 2007. Accessed February 4, 2012. http://content.time.com/time/specials/2007/article/0,28804,1686204_1686252_1690920,00.html.

Mangels, John. "Ancient Clams Discovered by College of Wooster Geologist Reveal How Earth Rebounded from Mass Extinction." *Cleveland Plain Dealer*, February 27, 2010. Accessed October 14, 2013. http://www.cleveland.com/science/index.ssf/2010/02/ancient_clams_discovered_by_co.html.

Marine Biological Laboratory. "Clams: They're Not Just for Chowder Anymore." August 12, 2004. Accessed July 26, 2014. http://hermes. mbl.edu/news/press_releases/2004/2004_pr_8_12d.html.

Melville, Herman. *Moby Dick; or, The Whale.* London, England: Richard Bently, 1851.

National Geographic Society. "Giant Clam—*Tridacna gigas.*" Accessed February 4, 2012. http://animals.nationalgeographic.com/animals/ invertebrates/giant-clam.

National Geographic Society. "Sustainable Seafood." Accessed November 24, 2013. http://ocean.nationalgeographic.com/ocean/take-action/ sustainable-seafood.

National Institute of Neurological Disorders and Stroke. "Brain Basics: Know Your Brain." Last modified April 28, 2014. Accessed October 14, 2013. http://www.ninds.nih.gov/disorders/brain_basics/know_your_ brain.htm

National Oceanic and Atmospheric Administration. "Ocean Quahog, *Arctica islandia*, Life History and Habitat Characteristics." NOAA Technical memorandum NMFS-NE-148, September 1999.

O'Hanlon, Larry. "Ancient Clams Lived 120 Years." Discovery.com. Last modified October 14, 2004. Accessed March 18, 2012. http://www. discovery.com/news/briefs/20041011/clams.html (site discontinued).

Oregon Department of Fish & Wildlife. "Life History of the Pacific Razor Clam." Accessed September 28, 2013. http://www.dfw.state.or.us/ MRP/shellfish/razorclams/lifehistory.asp.

———. "Bay Clam Identification." Accessed September 28, 2013. http:// www.dfw.state.or.us/MRP/shellfish/bayclams/ClamID.asp.

Passy, Charles. "Fish by Mail: Worth the Clams?" *Wall Street Journal,* August 13, 2009. Accessed October 11, 2013. http://online.wsj.com/ news/articles/SB10001424052970203496804574346393250601998.

Pike, Sue. "Clams: The Wonder of Their Reproduction." *Seacoast Online,* June 3, 2009. Accessed 3/17/2012. http://www.seacoastonline.com/ articles/20090603-LIFE-906030327.

Prince William Sound Regional Citizen's Advisory Council. "Non-Indigenous Aquatic Species of Concern for Alaska." Fact Sheet 4. Updated May 4, 2004.

Reaske, Christopher. *The Compleat Clammer: A Guide to Gathering and Preparation.* Short Hills, NJ: Burford Books, 1999.

Ridgway, Iain. "The Long-Lived Clam *Artica islandica*, A New Model Species for Ageing Research." British Society for Research on Ageing. Last modified June 23, 2008. Accessed February 4,

2012. http://www.bsra.org.uk/e-lifespan/my-favorite-system/long-lived-clam-arctica-islandica-new-model-species-ageing-research.

Ridgway, Merce. *The Bayman: A Life on Barnegat Bay*. Harvey Cedars, NJ: Down the Shore Publishing, 2000.

Roach, Mary. *Bonk: The Curious Coupling of Science and Sex*. New York: W.W. Norton, 2008.

Roegner, G. Curtis and Roger Mann. "Hard Clam: *Mercenaria mercenaria*." Gloucester Point, VA: Virginia Institute of Marine Science, (n.d.).

Rosauer, Ruth, ed. "Ancient Hawaiian Aquaculture." Rodale Institute. Accessed October 21, 2013. http://www.fadr.msu.ru/rodale/agsieve/txt/vol5/3/art8.html.

Rubenstein, Chanah. "Can Your First Name Hurt Your Chances in Online Dating?" *Digital Journal*, January 6, 2012. Accessed January 7, 2014. http://digitaljournal.com/print/article/317419.

Sacred Sites International Foundation. "The Emeryville Shellmound." Accessed December 2, 2013. http://www.sacred-sites.org/preservation/more_shell.html.

Science Daily. "50-Million-Year-Old Clam Shells Provide Indications of Future of El Nino Phenomenon." Last modified September 20, 2001. Accessed March 18, 2012. http://www.sciencedaily.com/releases/2011/09/110919113851.htm.

———. "There's Something Fishy about Human Brain Evolution." Last modified February 22, 2006. Accessed October 14, 2013. http://www.sciencedaily.com/releases/2006/02/060221090456.htm.

Seafood Health Facts. "Seafood Choices." Accessed October 12, 2013. http://seafoodhealthfacts.org/seafood_choices/clams.php.

Shimek, Ronald L. "Greetings, Clamrades." Reefs.org. Last updated May 1997. Accessed March 21, 2012. http://www.reefs.org/library/aquarium_net/0597/0597_1.html.

———. "Phytoplankton: A Necessity for Clams." DT's Live Marine Phytoplankton. Accessed September 27, 2013. http://www.dtplankton.com/articles/necessity.html.

Sifton, Sam. "Slurping Seashells." *New York Times*, July 13, 2010. Accessed November 8, 2013. http://www.nytimes.com/2010/07/14/dining/14note.html.

South Carolina Department of Natural Resources. Marine Resources Research Institute. *A Review of the Potential Impacts of Mechanical Harvesting on Subtidal and Intertidal Shellfish Resources*. By Loren D.

Coen. January 12, 1995. Accessed November 24, 2013. https://www. dnr.sc.gov/marine/mrri/shellfish/harvester.pdf.

Ste. Claire, Dana. "Volusia County's Rich Prehistoric Past." Volusia County Florida. Accessed November 22, 2013. http://www.volusia.org/residents/history/volusia-stories/people-of-the-shellmounds/volusia-countys-rich-historic-past.

Stein, Julie K. "The Analysis of Shell Middens" in *Deciphering a Shell Midden*, ed. Julie K. Stein. Philadelphia, PA: Academic Press, 1992.

Stein, Natalie. "Nutritional Value of Steamed Clams." *SFGate*. Accessed September 30, 2013. http://healthyeating.sfgate.com/nutritional-value-steamed-clams-1517.html.

Stradley, Linda. "History of Chowder." What's Cooking America. Accessed January 23, 2012. http://whatscookingamerica.net/History/ChowderHistory.htm.

Swan, Christopher. "Where to Find Rudistids in New York." *Christian Science Monitor*, August 28, 1980. Accessed December 6, 2013. http://www.csmonitor.com/1980/0828/082849.html.

Sydney Morning Herald. "Pearly Wisdom: Oysters Are an Aphrodisiac." March 24, 2005. Accessed September 30, 2013. http://www.smh.com.au/articles/2005/03/23/1111525227607.html.

Tarnowski, Mitchell. "Hard Shell Clam." Maryland.gov. Accessed 9/30/2013. http://dnr2.maryland.gov/Fisheries/Pages/Fish-Facts.aspx?fishname=Shellfish%20-%20Hard%20Shell%20Clam

University of Delaware. "Non-invasive Technique Reveals 'Ancient Harvests,' Dotting Delaware Shore." University of Delaware Department of Geology. Last modified March 25, 1998. Accessed March 19, 2012. http://www.udel.edu/geology/AncientHarvests.html.

Urban, Mike. *Clam Shacks: The Ultimate Guide to New England's Most Fantastic Seafood Eateries*. Kennebunkport, ME: Cider Mill Press, 2011.

Washington Department of Fish & Wildlife. "Butter Clams." Accessed September 28, 2013. http://wdfw.wa.gov/fishing/shellfish/clams/butter_clams.html.

WebMD. "Aphrodisiacs: Better Sex or Just Bunk?" Accessed October 15, 2013. http://www.webmd.com/sex-relationships/features/want-better-sex?

Welsh, Craig. "China's Demands for Geoducks Sends Prices, Profits Soaring in NW." *Seattle Times*, April 21, 2012. Accessed October 15, 2013. http://seattletimes.com/html/localnews/2018041537_geoduck22m.html.

Whetstone, Jack, L.N. Sturmer, and M.J. Oesterling. "Biology and Culture of the Hard Clam." Southern Regional Aquaculture Center: SRAC Publication No. 433 (August 2005 – revision).

White, Jasper. *50 Chowders: One-Pot Meals – Clam, Corn & Beyond*. New York: Scribner, 2000.

Winkler, Sarah. "How Aquaculture Works." How Stuff Works. Accessed October 18, 2013. http://animals.howstuffworks.com/animal-facts/aquaculture.htm.

Wyer, Holly. "Atlantic Surfclam, Ocean Quahog, Hard Clam and Softshell Clam." Monterey Bay Aquarium Seafood Watch. Monterey, CA: Monterey Bay Aquarium, 2012.

Young, Jim. "Clams." Netarts Bay Today. Accessed October 19, 2013. http://www.netartsbaytoday.org/html/clams_.html.

Zamora, Cherry. "The Rudists." University of California Museum of Paleontology. Accessed December 6, 2013. http://www.ucmp.berkeley.edu/taxa/inverts/mollusca/rudists.php.

Photo Credits

x: Photo by RonPaul86, http://commons.wikimedia.org/wiki/File:
Newport_Pier.JPG, http://creativecommons.org/licenses/by-sa/3.0/
legalcode, Date: 5/13/14

xii: Photo by Mary P. Madigan, "Clam Broth House", www.flickr.com/
photos/marypmadigan/236516077, Date: 4/11/14

12:   Photo by Mark A. Wilson, "Rudists", http://commons.wikimedia.
org/wiki/File:RudistCretaceousUAE.jpg, Date: 4/11/14

24:   Photo © Tom Carlisle, Compass Points Photography, All Rights
Reserved. "Shell Midden" http://commons.wikimedia.org/wiki/
File:Shell_middens_at_Grand_Bay_(5734564343).jpg, Date: 4/11/14

32:   Photo by author

34:   Photo by author

48:   Photo by Hannes Grobe/AWI, *Tridacna gigas - giant clam from the
Philippines (length 55 cm)*, http://commons.wikimedia.org/wiki/
File:Tridacna-gigas_hg.jpg, Date: 5/12/14

54:   Photo © Jean-Marie Hullot, *Avalokitessvara*, http://www.fotopedia.
com/items/jmhullot-jzjqXAGmm8g#context=de17d03e683, Date:
5/12/14

56:   Photo © Adriano C. *Portrait of Giacomo Casanova*, http://commons.
wikimedia.org/wiki/File:Casanova_ritratto.jpg Date: 5/12/14

71:   Painting by Fra' Filippo Lippi, *Madonna and Child*, http://commons.
wikimedia.org/wiki/File:Madonna_and_Child_(Filippo_Lippi).
jpg#mediaviewer/Archivo:Madonna_and_Child_%28Filippo_
Lippi%29.jpg Date: 5/13/14

72:   Painting by Botticelli, *The Birth of Venus*, http://commons.wikimedia.
org/wiki/File:La_nascita_di_Venere_(Botticelli).jpg, Date: 5/13/14

74: Painting by Winslow Homer, *A Basket of Clams*. Photograph by Metropolitan Museum of Art, http://commons.wikimedia.org/wiki/File:Winslow_Homer_-_A_Basket_of_Clams.jpg, Date: 5/13/14

75: Painting by William James Glackens, *Treading Clams at Wickford*, http://www.wikipaintings.org/en/william-james-glackens/treading-clams-wickford-1909, http://www.the-athenaeum.org/art/detail.php?ID=33152 Date: 5/13/14

89: Photo by author

94: Photo by Jessica Spengler, *Razor Clams,* http://commons.wikimedia.org/wiki/File:Razor_clams_by_H.R._Giger_(8053207185).jpg, Date: 5/13/14

97: Photo by Paul Keleher, *Soft Shell Clams*, http://commons.wikimedia.org/wiki/File:Steamedclams.jpg, Date: 5/13/14

106: *mollusk_diagram* Credit: http://coolidgelifesciencetext.wikispaces.com/3g.+Mollusks, http://creativecommons.org/licenses/by-sa/3.0/legalcode, Date: 5/13/14

125: Photo by Gordon T. Taylor, http://en.wikipedia.org/wiki/File:Diatoms_through_the_microscope.jpg, Date: 5/13/14

155: Photo by author

157: Photo by author

158: Photo by author

159: Photo by author

161: Photo by Starest Westst, *Geoducks*, http://commons.wikimedia.org/wiki/File:HK_seafood_Elephant_Trunk_Clam.jpg, http://creativecommons.org/licenses/by-sa/3.0/legalcode, Date: 5/14/14

163: Photo by author

167: Photo by Collin Grady, *Alekoko "Menehune" fishpond*, http://en.wikipedia.org/wiki/File:Alekoko_fishpond.jpg, http://creativecommons.org/licenses/by-sa/2.0/legalcode, Date: 5/14/14

174: Photo by author

176: Photo by author

177: Photo by author

179: Photo by author

180: Photo by author

190: Photo by author

192: Illustration by Chiswick Chap, http://en.wikipedia.org/wiki/File:Tides_overview.png, http://creativecommons.org/licenses/by-sa/3.0/legalcode, Date: 5/15/14

197: Photo by author

199: Photo by author
201: Photo by author
205: Photo by author
206: Photo by author
229: Photo by author

Index